Lecture Notes in Nanoscale Science and Technology

Volume 24

Lecture Notes in Nanoscale Science and Technology (LNNST) aims to report latest developments in nanoscale science and technology research and teaching – quickly, informally and at a high level. Through publication, LNNST commits to serve the open communication of scientific and technological advances in the creation and use of objects at the nanometer scale, crossing the boundaries of physics, materials science, biology, chemistry, and engineering. Certainly, while historically the mysteries in each of the sciences have been very different, they have all required a relentless step-by-step pursuit to uncover the answer to a challenging scientific question, but recently many of the answers have brought questions that lie at the boundaries between the life sciences and the physical sciences and between what is fundamental and what is application. This is no accident since recent research in the physical and life sciences have each independently cut a path to the edge of their disciplines. As both paths intersect one may ask if transport of material in a cell is biology or is it physics? This intersection of curiosity makes us realize that nanoscience and technology crosses many if not all disciplines. It is this market that the proposed series of lecture notes targets.

Virat Khanna • Chander Prakash

Graphene for Electrochemical Energy Storage

Energizing the Future

Springer

Virat Khanna
University Centre for Research and Development
Chandigarh University
Gharuan, Mohali, Punjab, India

Chander Prakash
Research and Innovation Cell
Rayat Bahra University
Mohali, Punjab, India

ISSN 2195-2159 ISSN 2195-2167 (electronic)
Lecture Notes in Nanoscale Science and Technology
ISBN 978-3-031-87642-4 ISBN 978-3-031-87643-1 (eBook)
https://doi.org/10.1007/978-3-031-87643-1

This Springer imprint is published by the registered company Springer Nature Switzerland AG
The registered company address is: Gewerbestrasse 11, 6330 Cham, Switzerland

Preface

Graphene, a two-dimensional carbon allotrope, has gained considerable attention due to its remarkable physical and chemical characteristics. It is considered one of the most promising materials for various applications such as electrochemical energy storage due to its large surface area, high electrical conductivity, and superior mechanical strength. Graphene could potentially serve as an ideal material for developing high-performance electrochemical energy storage devices.

This book provides a comprehensive overview of graphene-based electrochemical energy storage applications. It examines the fundamental properties of graphene, including its structure, electronic characteristics, and synthesis methods. Furthermore, the book highlights various electrochemical energy storage applications for graphene such as supercapacitors, batteries, and fuel cells. The initial section of this book provides a comprehensive introduction to graphene and its properties, such as electronic, mechanical, and thermal ones. It discusses various synthesis methods for graphene such as mechanical exfoliation, chemical vapour deposition, and epitaxial growth. The second chapter of this book discusses the fictionalization of graphene which could enhance its characteristics for electrochemical energy storage applications. Specifically, it discusses its use in supercapacitors—devices used as energy storage devices when high power density is needed. It also discusses different types of supercapacitors such as electrochemical capacitors, pseudocapacitors, and hybrid capacitors and how graphene can enhance their performance. In the third chapter of the book, the role of graphene is explored in batteries. By using graphene-based anodes and cathodes in batteries, they can improve their energy density, cycling stability, and rate capability—across various types including lithium-ion, sodium-ion, and zinc-ion batteries. In the concluding chapter of this book, the role of graphene is explored in fuel cells—electrochemical devices that transform chemical energy into electrical power. Graphene-based materials can be utilized as catalysts or supports within fuel cells for improved performance and durability. This chapter examines various types of fuel cells such as proton exchange membrane fuel cells, direct methanol fuel cells, and solid oxide fuel cells.

This book will also highlight the prospects and challenges in graphene-based electrochemical energy storage applications. This book is a great reference for researchers, scientists, and engineers working in this area of energy storage involving graphene as the primary material.

Gharuan, Mohali, Punjab, India Virat Khanna
Mohali, Punjab, India Chander Prakash

Contents

About the Authors

Virat Khanna works as an Associate Professor at the UCRD, Chandigarh University, Punjab, India. He has more than 15 years of experience in academics and research. He has authored over 50 national and international publications in SCI, Scopus, and Web of Science-indexed journals. He is the editor of several books and conference proceedings with international publishers like Springer, CRC Press, IGI Global, and Bentham Science Books. He has delivered several invited seminars and keynote talks on international platforms and has been awarded the best research paper awards. He also serves as the Editorial Board member of the prestigious journal *Scientific Reports* of Nature Group and is also the academic editor of the *Journal of Nanotechnology*, Wiley and *Advanced Materials Science and Engineering*, Wiley. He is also the Book Series Editor entitled *Recent Advances in Materials for Nanotechnology* of CRC Press, Taylor & Francis Publisher. He has also guest edited various special issues with various international journal publishers like Springer, De Gruyter, etc. He is also a reviewer board member of several international journals.

Chander Prakash serving as Group Pro-Vice Chancellor, Rayat Bahra University, Mohali, Punjab, India. He served as Dean of Research and Development/School of Mechanical Engineering at Lovely Professional University, India. He did his PhD in Mechanical Engineering at Panjab University, Chandigarh. He has published over 530 scientific articles in peer-reviewed journals. He is a highly cited researcher with 13400 citations (H-index 66 and I-10 index 252: Google Scholar 2025). He is listed in the Top 2% world's leading scientists published by Stanford University and Elsevier. Dr. Prakash edited/authored 30 books, serving as Series Editor of 3 books, as Guest Editor and serving as Editorial Board Member of Journals. He has published 18 patents. Dr. Prakash raised funds for Mega Grants in collaboration with Southern Federal University, Russian Federation.

Chapter 1
Introduction to Graphene

1.1 Introduction

Science and engineering have been progressing persistently in the field of materials. The advancement in materials technology can be traced back from its origin to humans understanding and interpretation of materials systems. Since the dawn of human civilisation, materials have become an essential part of life. Several new advanced materials have been developed and produced over time, and research is being carried out constantly to generate new and cutting-edge materials with improved properties. It is often considered that the behaviour of a material depends completely on the material's composition. However, the size, configuration, and dimensionality also have a significant effect on how the material will behave. Among all the available materials, nanomaterials have gained special interest due to their exceptional physicochemical properties, such as nanomaterials exhibit super-magnetism.

1.2 Evolution of 2D Materials

Two-dimensional (2D) materials are fundamentally crystalline materials produced from a single or few layers of atoms. Several decades ago, the scientific community had a view that 2D materials are unstable and cannot be synthesised [1, 2]. However, so far, numerous 2D materials have been theoretically identified and developed. At the same time, many more are yet to be formed. A large group of them belongs to the carbon family, such as graphite. Since the sixteenth century, graphite has emerged as a prominent material for innumerable applications, including dry lubricants, crucibles, and foundry facings, as well as steel production. In 1962, during the period of persistent research on carbon, Boehm introduced the

V. Khanna, C. Prakash, *Graphene for Electrochemical Energy Storage*, Lecture Notes in Nanoscale Science and Technology 24,
https://doi.org/10.1007/978-3-031-87643-1_1

suffix "ene" to the chemically unstable allotropic form of carbon. Since the discovery of 0D fullerenes in 1985, several additional carbon allotropes have emerged [3]. This confirmed the existence of 1D carbon nanotubes (CNTs) as well. The development of CNTs in 1991 by Iijima revolutionised the world of materials [4]. This prompted researchers to carry out further investigation to create 1D nanoribbons or segregate 2D graphitic materials from 2D crystals [5]. Since then, several findings on fullerene, graphite, and CNTs have been documented, with graphite being considered to be the original material. However, an absolute transformation in the area of 2D materials occurred after 2004. When, for the first time, a single layer of graphene sheet was extracted by the scientists. Graphene, which is composed of carbon atoms arranged in a honeycomb-like hexagonal structure, has intriguing features that set it apart from graphite. Graphene, the thinnest substance known, is the strongest material yet measured, measuring just one atom thick. Despite being a cutting-edge substance, its graphitic layered structure was first observed back in 1859 by English Chemist Benjamin Collins Brodie [6]. Hanns-Peter Boehm first used the name graphene in 1962 to describe extremely thin graphite flakes in a theoretical monolayer structural analysis [7]. However, Dreyer and his team noted that Boehm's work was based on heteroatomic contamination of isolating reduced graphene oxide rather than pure graphene [8]. From 1990 to 2004, several efforts were put forward using complex systems to create immensely thin layers of graphite. However, nothing less than 50–100 layers thick was formed during this time [9].

1.3 Discovery of Graphene

Based on theoretical research on the thermodynamic stability of 2D crystals, the isolation of single-layer graphene appeared to be unattainable for many decades [10]. However, in 2004, an unalike and naive approach was applied by the research team in Manchester led by Geim and Novoselov to segregate graphene from graphite layers. A micromechanical cleavage process was utilised to separate a monolayer of carbon atoms from a 3D graphite structure [11, 12]. During one of the Friday night experiments, Andre Geim and Kostya Novoselov used a sticky tape, due to shorter time availability, to remove flakes from a lump of bulky graphite to get insights about its physical behaviour. It was observed that some of the removed flakes were comparatively thinner than others. After several rounds, nearly 10 times of segregating the graphite flakes, a new wondrous material, graphene, was extracted from weakly coupled crystals, and it became possible to examine the physical properties of the first carbon-based 2D material [13, 14]. The researchers thus avoided all the complications with the stability of small crystallites by employing this top-down strategy using three-dimensional crystals.

1.4 Structure of Graphene

Carbon is one of the Periodic Table's most fascinating elements [15]. Exceptional properties of carbon empower it to be arranged into innumerable structures, including fullerene (0D), nanotube (1D), graphene (2D), and diamond (3D) [16]. The carbon atom has six electrons, two of which are located in the $1s^2$ shell and four of which are located in the 2s and 2p states with a ground-state atomic configuration of $2s^2 2p^2$ [17, 18]. The electrons in a carbon atom's valence shell can form three different types of hybridisation, namely sp, sp^2, and sp^3, providing linear, planer, and 3D structures [19, 20]. Particularly exciting are sp^2 hybridisations. As shown in Fig. 1.1a, in sp^2 hybridisation orbitals, a triangle is formed in one plane while the p_z orbital remains in a plane which is perpendicular to it [21]. The π-bond is formed by the p_z electron. To establish the π–π bands, the π-bonds hybridise with one another [22]. Graphene, which serves as the foundation for the vast majority of graphitic materials in its ideal form, is mainly composed of a sp^2 hybridised single sheet of carbon atoms arranged in a hexagonal structure [23–25].

Figure 1.1 depicts the hexagonal carbon ring structure of graphene with an area of 0.052 nm^2 and carbon-to-carbon bond length of 0.142 nm. In a single cell of graphene, represented by a red rhombus, a lattice constant of 2.46 Å is shared by the unit-cell vectors a1 and a2 [26]. In addition, graphene possesses a resultant covalent sigma bond in a 2D plane having the shortest interatomic distance of 1.42 Å, which in turn makes it even stronger than a sp^3 hybridised carbon-to-carbon bonding in diamond [27]. As a result, the mechanical properties of single-layer graphene improve substantially. Further, the partially filled π band also results in a zero-band gap between the valance band and conductive band, which eventually causes free-moving electrons in single-layer graphene. Apart from this, due to weak π-bonds, poor van der Waals reaction arises among the adjoining layers in both bilayer and multilayer graphene [26, 27]. Graphene is theoretically 0.34 nm thick, which is approximately 1/200,000th the diameter of human hair [28]. Even though graphene

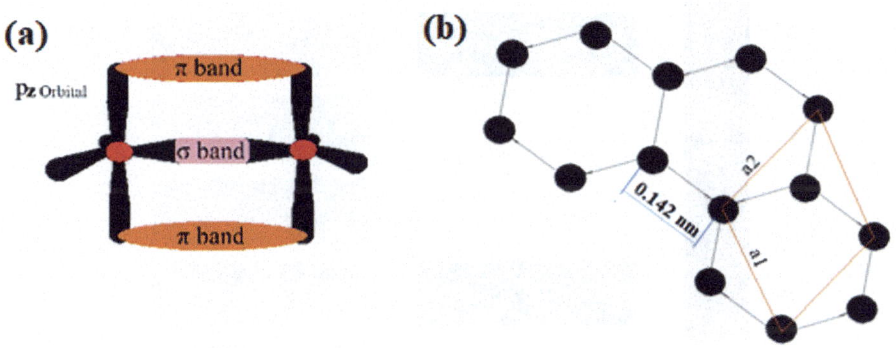

Fig. 1.1 Shows (**a**) sp^2 hybridised orbital structure and (**b**) hexagonal structure of graphene

is the thinnest material, its structure is significantly stable [29]. To deter the reformations of atoms, the bond among the carbon atoms is strong enough to withstand the externally applied forces induced through a lattice-twisting plane [22].

1.5 Synthesis Methods

Graphene has been gaining massive attention around the world for its unmatched properties and great potential in distinct areas. Owing to its abilities, there have been some rapid advancements in graphene synthesis approaches. As shown in Fig. 1.2, the synthesis of graphene is broadly categorised into two most common approaches. The top-down method is employed, separating the graphene layers from a bulk of graphite. At the same time, hydrocarbon substances as precursors are used to build carbon blocks of graphene in the bottom-up method [30].

1.5.1 Exfoliation Method

Mechanical exfoliation is one of the simplest and primary synthesis methods to produce high-quality graphene sheets. This technique consists of using directional force, normal or lateral, to the bulk graphite to peel layers of graphene by reducing the effect of van der Waals forces between the stacked layers [31]. Mechanical exfoliation methods are further classified as micromechanical cleavage, continuous mechanical cleavage, and shear exfoliation methods. These methods involve the use of mechanical force using tapes, mills, and blenders.

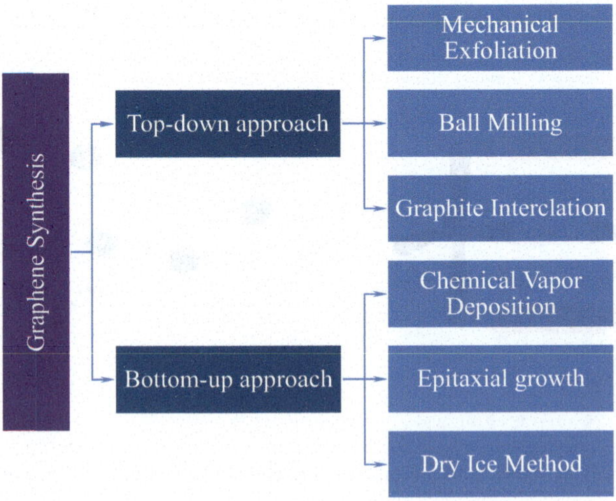

Fig. 1.2 Routes for graphene synthesis

1.5.1.1 Micromechanical Cleavage

A single layer or multiple layers of graphene can be obtained using the mechanical stripping process, which presses a sheet of highly oriented pyrolytic graphite (HOPG) or natural graphite on the target surface and repeatedly peels it off [32–35]. Using a scotch tape, Novoselov and Geim were able to obtain monolayer graphene for the first time using this technique in 2004 [14]. Several other approaches, including ultrasonication, electric field, and printing, can also be utilised for peeling off the graphene sheets [36–39]. Besides this, to increase the production rate of graphene flakes, HOPG has also been coupled to the substrate using adhesives, such as resin [36, 40]. Micromechanical exfoliation is an inexpensive method to synthesise high-quality graphene. However, the non-uniform thickness of sheets and the inability to produce graphene at a large scale are major drawbacks of this method [41].

1.5.1.2 Continuous Mechanical Cleavage

Another advanced version of the Scotch tape method for the production of single-layer and few-layer graphene from raw graphite is the roll-milled technique [42]. In this method, graphite and polymer adhesive, often polyvinyl chloride (PVC), are sandwiched in between the rotating rolls. This allows graphite exfoliation through shear force to produce graphene. Moreover, producing graphene in a pure state without any adhesives for commercial purposes is essential [43]. The removal of adhesives, such as PVC, is not only complex and energy-intensive but also takes prolonged hours of heating at high-temperature ranges [42].

1.5.1.3 Shear Exfoliation

Exfoliation of graphene may also be carried out by applying high shear force to a graphite in a stabilising liquid. Sodium cholate is usually used as a stabilising liquid to perform the shear exfoliation of the precursor. This approach to exfoliate graphite has been highly useful to achieve a shear rate of more than 10^{-4}/s, even with a kitchen blender [44]. Even at extremely high shear rates, this process's disintegration output is as minimal as 0.1%, notwithstanding how easy it is to scale up. It has been noted that to boost the scalability of the process, aqueous solvents can be used instead of organic solvents. Advancing the system by attaching fans to control the process temperature can also have a significant effect in improving its efficacy [45].

1.5.2 Ball Milling

Ball Milling (BM) is one of the most widely used processes, particularly in manufacturing industries. Several developments have been made in this process to improve the efficiency of the process, which is why it is important to understand it separately. Fundamentally, BM is used for blending materials and to reduce the size of the materials. Different techniques, including chemical or solvent assistance, have been used to separate graphene flakes from graphite [46, 47]. Chemical interactions restrict the sheets from restacking. Magnetic ball milling assists in the filtration of metallic compounds which may be present in the precursors [48]. With such methods, great opportunities arise to synthesise few-layer graphene and high yield. Also, graphene obtained can have a higher surface area of 153.9 m^2/g in comparison with graphite, along with an exceptionally low (0.66) I_D/I_G ratio [49, 50].

1.5.3 Graphite Intercalation

Graphite intercalation is another useful method to produce graphene. There are two techniques to carry out graphite intercalation. Firstly, through introducing tiny molecules into the bulk graphite layers. Secondly, associating molecules or polymers to the graphite sheets using non-covalent bonds which leads to the formation of graphite intercalation compounds (GICs) [51]. Using sulphuric acid and hydrogen peroxide as intercalant and oxidants, the spheroidised graphite was transformed into graphene nanoplatelets effectively [52]. Similarly, the sound transmissibility is also enhanced with polymer intercalation [53]. During intercalation in GICs, the molecules are placed at the interface galleries while the graphite layers remain intact [54]. Also, it is important to note that GICs produced with different intercalants can have different effects on the properties of the graphene [51, 55].

1.5.4 Chemical Vapour Deposition

In CVD, organic gaseous reactants are deposited onto the metallic substrate to produce graphene. A transitional metal catalyst is initially used to adsorb the carbon precursor in the gas phase onto its surface. The catalyst then proceeds through catalytic disintegration to produce carbon atoms. At high temperatures, the carbon compounds that have been disintegrated eventually diffuse, dissolute, precipitate, and deposit onto the surface of metallic substrates. Surface nucleation begins during cooling, followed by the development of graphene [56]. CVD is regarded as a highly promising technique for synthesising graphene with high quality and larger surface area, in particular with Ni and Co having an incomplete shell. This makes graphene extremely suitable for developing graphene films [57]. CVD is further divided into two groups, namely thermal CVD and plasma CVD. The difference between them lies in the processing temperature for graphene growth. In contrast to thermal CVD, plasma-enhanced CVD (PEVD) provides a different mean for producing graphene at comparatively lower temperature value [48]. Added innovation in PECVD has led the researchers to reduce the growth temperature further. Microwave PECVD can be executed at a much-reduced temperature of 760 °C, which permits swift processing in one step [58]. Further, Ni-Cu alloys have also been used to improve the CDR and catalytic activity for the growth of graphene on SiO_2/Si substrate at low temperature [59]. Similarly, with an Ni catalyst, enhanced CVD can synthesise highly crystallised several-layer graphene at 464 °C temperature [60].

1.5.5 *Epitaxial Growth*

One of the most famous processes for producing graphene is epitaxial thermal deposition on a surface of silicon carbide (SiC) substrate at 1100 °C. In other words, epitaxial growth is a process in which an epitaxial film of graphene is produced by depositing a single crystalline layer on an SiC substrate's single crystal [61]. However, graphene's fake size will rely on the SiC wafer's size. Carrier density, mobility, and material thickness are also influenced by the substrate surface [62]. There are two types of epitaxial layers, namely the homo-epitaxial layer and the hetero-epitaxial layer. Homo-epitaxial layers are those in which the deposited film and the substrate are of the same material, while in the hetero-epitaxial layer both are different [63]. Epitaxial growth of high-quality graphene on SiC substrate has great potential although it is an expensive process as SiC substrate is costly and the yield obtained is low. Therefore, this method is unsuitable for commercial processing.

Another way of epitaxial growth of graphene sheets is from metal-carbon melts. It is a straightforward process which involves locating a carbon source, generally graphite powder, along the transition metal and melting the metal at an elevated temperature [48]. However, the temperature can be reduced to reduce carbon's solubility once it begins to mix in the liquid metal. This, however, may cause an excessive amount of carbon to precipitate. Different types of carbon allotropes, such as thick graphite, few-layer graphene (FLG), and monolayer graphene, can be obtained by skimming off the precipitate. Nickel (Ni) is considered the most suitable metal for this process as it is not Raman active [64]. Other metals, such as ruthenium and iridium, may also be used for graphene production, but they may produce low-quality graphene as compared to nickel or copper (Cu) [65, 66]. For example, under a controlled environment using argon as an inert gas, carbon from methane gas deposits onto the nickel film, which had been heated at high temperature, and forms graphene sheets [67, 68].

1.5.6 *Dry Ice Method*

The dry ice method allows direct conversion of carbon dioxide (CO_2) into thin, few-layer graphene. It has been observed that burning magnesium metal in a CO_2 environment produces graphene resulting from a high exothermic reaction. The residue, after complete burning, is stirred for a prolonged time in an HCl solution. Once both

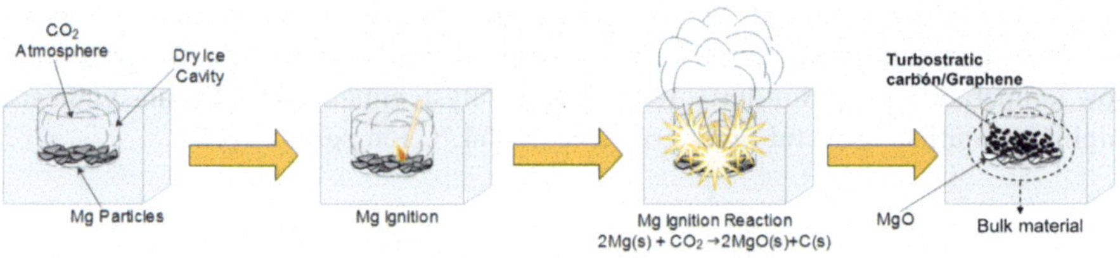

Fig. 1.3 Schematic of synthesising graphene through dry ice method [70]

the Mg and MgO are dissolved in water, their mixture is filtered with water to remove the impurities. After drying in a vacuum, graphene is obtained [69]. Similarly, turbostatic carbon/graphene can also be obtained by burning Mg chips in a CO_2 atmosphere, as shown in Fig. 1.3 [70]. Besides this, to enhance the processing efficacy, along with Mg, calcium (Ca) can be used to enhance the electrochemical characteristics of graphene-based carbon supercapacitors [71]. There have been several advantages associated with this technique, such as the processing cost is quite low, a far higher yield of graphene is possible compared to other conventional methods, and the process is eco-friendly as purifying water can be neutralised immediately. However, this method is also not without drawbacks. Significant oxide amounts cannot be eradicated during the cleaning process, which may not be suitable for further usage.

1.6 Various Characterisation Techniques of Graphene

Graphene is regarded as the mother of all allotropic forms of carbon. There are several derivatives of graphene. At the same time, the most common among them are graphene oxide (GO, reduced graphene oxide (rGO), graphene nanoplatelets (GNPs), graphene nanoribbons (GNRs), graphene nanosheets (GNS), and few-layer graphene (FLG) [72–76]. To examine the surface functionality, chemical structure, and properties of graphene, numerous new and significant analysis testing methods have been developed. Graphene characterisation investigations involve the study of structural images, such as optical microscopy, transmission electron microscopy (TEM), scanning electron microscopy (SEM), and graphs like Raman spectra and X-ray diffraction (XRD).

1.6.1 Optical Microscopy

In optical microscopy, graphene is observed closely with the use of a lens magnifier and visible light. This is a conventional technique which is still widely used to understand the structural behaviour of material. It is a common, inexpensive method

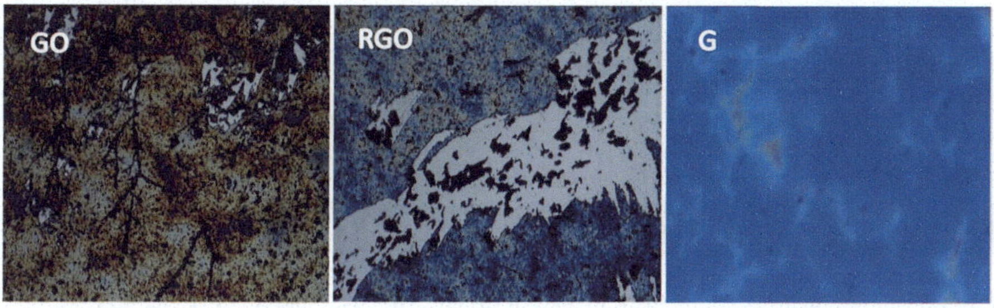

Fig. 1.4 Shows the optical lens microscopic images of graphene and its derivatives in powder form [79]

that provides fair imaging of graphene structure [77]. Figure 1.4 shows the optical microscopy of graphene and its derivatives. Graphene synthesised through distinct methods usually has great differences in morphology, structure, and layer numbers. For example, graphene oxide exhibited a lateral dimension of <5 μm and a thickness of 1 nm, while FLG had a thickness of 0.7–1.2 nm [78]. However, to have an in-depth understanding of graphene structural behaviour, often two or more imaging techniques are combined.

1.6.2 Scanning Electron Microscopy

SEM provides magnified images to observe the surfaces of a specimen or powder. Instead of light, electrons are used to form surface images with high resolution. The secondary electron from the sample surface is released when it is exposed to a focused electron beam (electron probe). Graphene sample topography can be viewed by 2D scanning of the electron beam, and an image can be obtained for analysis. SEM images of graphene derivatives are shown in Fig. 1.5. Compared to optical microscopy, SEM provides higher resolution and magnification. With sub-nanometer resolution, the most advanced SEMs may attain magnifications of approximately 100,000 times.

1.6.3 Telescopic Electron Microscopy

The TEM is a cutting-edge technology designed to analyse and visualise small-sized samples ranging from micro to nanoscale. This technique can disclose extremely intricate details that are unattainable using traditional optical light microscopes [83]. In this process, a beam of electrons from an electron gun penetrates through the material sample before it arrives at the image lens and receiver [84]. TEM provides higher resolution and determines the morphological structure at the

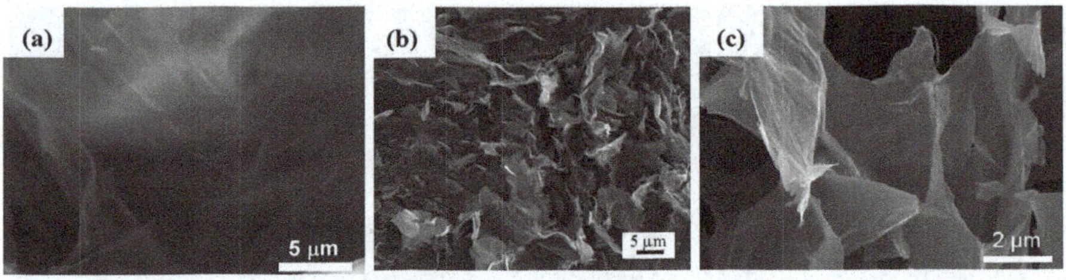

Fig. 1.5 Shows high-resolution SEM images of (**a**) GO, (**b**) GNS, and (**c**) GNP [80–82]

Fig. 1.6 Depicts X-ray diffraction pattern of graphite, graphene, and graphene oxide [85]

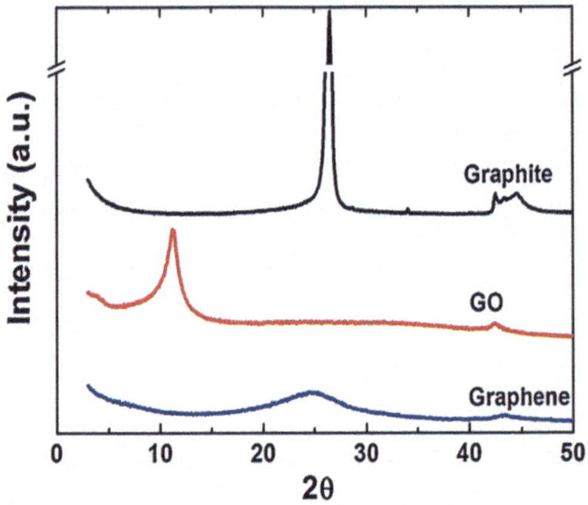

atomic scale of the graphene surface. As graphene and its derivatives, particularly rGO, are atomically thick layers, TEM has been observed as the single instrument that can exhibit the atomic behaviour of graphene. This method is relatively simple and consumes less time.

1.6.4 X-Ray Diffraction

XRD is regarded as the most significant technique to unravel all the problems about the crystalline structure of materials, such as crystallite size and morphology, the recognition of different substances, the orientation of crystals, and defects. During XRD, diffraction patterns are produced as a result of X-ray interactions with the material crystals. Thus, X-ray diffraction is often executed to determine the structure of graphene and its derivatives, as shown in Fig. 1.6. While graphite exhibited the highest peak at 27 °C, GO shows a characteristic peak at 11 °C. This elucidates that spacing has increased due to the presence of oxygen-containing groups, which increases the distance between the stacked layers of graphene [85].

1.7 Types and Derivatives of Graphene

The constant development in 2D materials has given rise to various types and derivatives of graphene with distinct properties and applications. These include single-layer graphene, bilayer graphene, few-layer graphene, and graphene oxide, each offering unique characteristics that have sparked significant interest in scientific research and technological advancements.

1.7.1 Single-Layer Graphene

In a single atomic-layer graphene sheet, the arrangement of atoms forms a honeycomb structure, where carbon atoms are bonded in a sp^2 configuration. The carbon-carbon bond length within the sheet is approximately 1.42 Å [86]. Also, it exhibits a linear and gapless band structure [87]. Single-layer graphene exhibits unique quantum effects, including the quantum Hall effect and ambipolar field effect [88]. In the presence of a strong magnetic field perpendicular to the graphene plane, the charge carriers in graphene, known as Dirac fermions, exhibit quantised Hall conductance [89]. This effect manifests as plateaus in the Hall resistance as a function of the magnetic field, indicating the quantisation of electronic states in graphene [89]. Also, due to the unique electronic structure of graphene, the application of an electric field can modify the carrier concentration in the material. In graphene, both electrons and holes can contribute to the current flow, making it ambipolar [90]. However, the deficiency (below 10 at room temperature) of an energy gap in the electronic structure of graphene results in a low on-off ratio for electronic transistors based on graphene [91]. For example, the lack of a band gap in monolayer graphene poses challenges for its application in transistor devices [92].

1.7.2 Bilayer Graphene

The stacking of two layers of graphene alters the electron energy dispersion and subsequently modifies its electronic properties due to the mutual interaction between the layers [93]. It is important to note that bilayer graphene should not be regarded as two separate layers of graphene but rather as a single composite electronic system [90]. In AB-stacked bilayer graphene, the carbon atoms of the top layer are positioned directly above the centre of the hexagon formed by the six-carbon ring of the bottom layer. The top layer's carbon atoms in AB-stacked bilayer graphene are positioned immediately above the centre of the hexagon created by the bottom layer's six-carbon ring. In this system, the conduction band with the lowest energy aligns with the valence band with the highest energy, resulting in a zero-band gap. Yet the symmetry can be disrupted by putting a vertical electric field between the two layers, which causes a non-zero band gap to appear. Depending on the strength of the electric field, it is possible to regulate the size of this induced band gap. The discovery of the induced band gap in AB-stacked bilayer graphene has significantly increased its appeal for electronic applications that depend on the presence of a band gap [94]. For instance, the dual-gated bilayer graphene Field-Effect Transistors (FETs) offer promising solutions, exhibiting high on/off ratios of up to 100 at room temperature. These bilayer graphene FETs hold the potential for the development of flexible and gate-tunable light-emitting devices and light detectors [92]. This development makes graphene a promising candidate for applications such as pseudospintronics and terahertz technology [95, 96].

1.7.3 Few-Layer Graphene

Few-layer graphene refers to graphene structures consisting of a small number of layers, as shown in Fig. 1.7, typically ranging from three to ten layers [97]. It retains some of the properties of single-layer graphene, such as high electrical conductivity

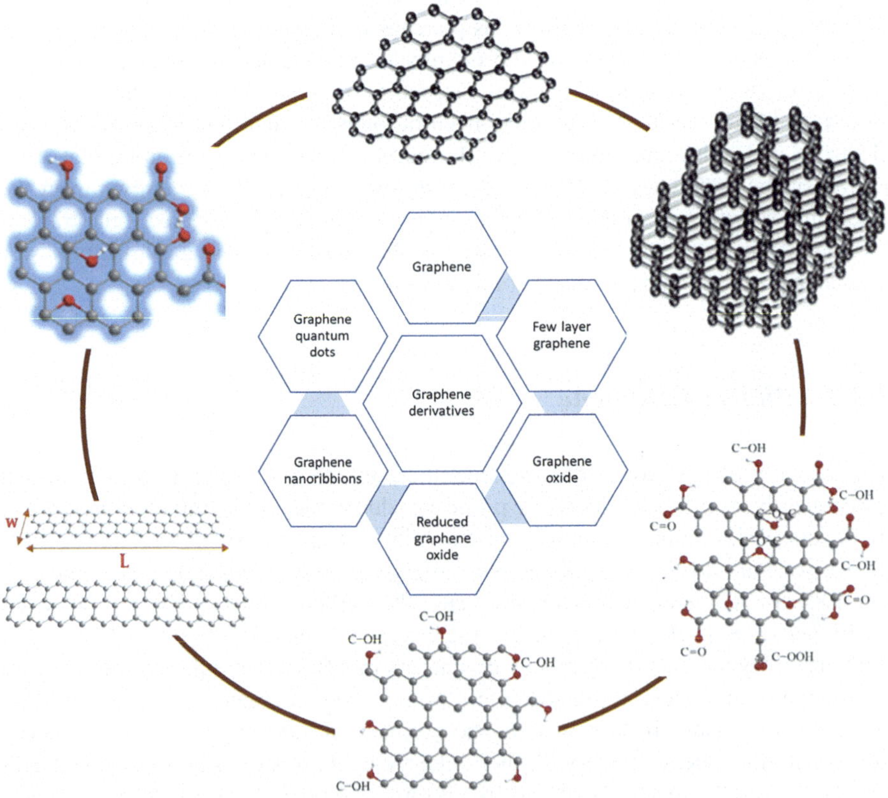

Fig. 1.7 A visual depiction illustrating the structure of graphene derivatives [112, 113]

and mechanical strength [97]. However, as the number of layers increases, interlayer interactions become more prominent, resulting in modifications of electronic properties [97]. FLG exhibits unique properties, such as a tunable band gap and enhanced light absorption, compared to single-layer graphene [98]. FLG can be synthesised using a variety of techniques, including microfluidisation, exfoliation, and CVD [99–101]. The number of layers and the stacking order of few-layer graphene can be precisely controlled, enabling the tailoring of its properties for specific applications.

1.7.4 Graphene Oxide

Graphene oxide is a derivative of graphene obtained through chemical processes that introduce oxygen-containing functional groups, such as epoxy, hydroxyl, carbonyl, and carboxylic, onto the graphene surface [102]. GO can also have either a single layer or a multilayer structure, like any 2D carbon material [103]. Initially, graphite is oxidised to form graphite oxide to synthesise GO, and subsequently, this graphite oxide is exfoliated to form GO. The properties of GO wholly rely on the oxidation technique, which decides the number and form of oxygen-containing groups in the resulting GO [104]. It has been evident that as oxidation increases, the number of oxygen-containing groups increases, decreasing the material's electrical properties as compared to highly reduced GO [105]. GO's dispersibility in water and other solvents makes it a versatile material that can be used in a variety of applications. For example, GO can be used to make composites, films, and coatings. It can also be used as a sensor or catalyst [106].

1.7.5 Reduced Graphene Oxide

One of the most appealing aspects of GO lies in its ability to undergo partial reduction, leading to the production of graphene-like materials, such as reduced graphene oxide (rGO). This process involves the removal of oxygen-containing groups, allowing the structure and properties of graphene to be restored to a certain degree [107]. The objective is to reduce GO to produce graphene-like sheets that closely resemble the pure graphene obtained from graphite, exhibiting similar characteristics in both structure and properties. The two most commonly employed strategies for the reduction of GO are thermal and chemical methods [108]. In the chemical approach, diverse reducing agents like hydrazine are utilised, albeit it poses toxicity concerns and results in the incorporation of nitrogen heteroatoms into the reduced GO [109]. rGO has several potential applications in electronics, energy storage, biomedical, catalysis, water purification, and composites [110]. It is a highly conductive material that can be tuned to have a variety of properties [111].

1.7.6 Graphene Nanoribbons

Graphene nanoribbons (GNRs) are thin graphene strips with a width of fewer than 50 mm [114]. The dimensions of graphene nanoribbons can be determined by measuring their length and width [115]. The performance of graphene nanoribbons (GNRs) is predominantly determined by their edge orientation. In many fields, the oxygenated derivative of GNRs has gathered significant attention due to the abundant presence of oxygen-based functional groups on both the edge and basal plane. These oxygen functionalities contribute to the unique properties and potential applications of oxygenated GNRs, particularly in biomedical research [116]. The aggregation tendency of graphene nanoribbons (GNRs) in solid or solution forms is primarily attributed to the π—π interaction between sp^2 hybridised carbon atoms. To mitigate this interaction and minimise aggregation, the introduction of functional substituents has been identified as an effective approach [117]. Several methods have been explored, including the introduction of alkyl substituents, functionalisation with high molecular weight polymers, and the incorporation of bulky 3D substituents on the surface of GNRs [118–120]. These strategies effectively reduce the π–π interaction and enhance the dispersibility of GNRs in solution. Additionally, graphene nanoribbons (GNRs) offer a means to modulate the electronic properties of graphene. Specifically, armchair GNRs (AGNRs) exhibit semiconductor characteristics, with their band gap diminishing as the ribbon width increases [121–123]. Conversely, zigzag GNRs (ZGNRs) experience an increase in band gap as the ribbon width expands. This tunability in electronic behaviour enables the precise control and engineering of graphene-based devices and systems [121–123].

1.7.7 Graphene Quantum Dots

Graphene quantum dots (GQDs) are small, single-layer graphene structures with dimensions in the nanometer range [124]. They exhibit unique electronic and optical properties due to quantum confinement effects, giving rise to their name as quantum dots [125]. GQDs are considered zero-dimensional entities, as their size is confined in all three dimensions [126–128]. The properties of GQDs can be tuned by varying the size, shape, and functionalisation of the particles [129]. For example, the size of the GQDs determines their fluorescence wavelength [130]. The shape of the GQDs affects their electrical properties [131].

There are two main approaches to synthesising graphene quantum dots (GQDs). The first approach is a top-down strategy, which involves cutting larger graphene sheets into smaller fragments to obtain GQDs [132]. This method utilises techniques such as chemical or laser etching, electrochemical exfoliation, or mechanical exfoliation to create GQDs from bulk graphene materials [133]. The second approach is a bottom-up strategy, where small precursor molecules, typically aromatic molecules, are used to generate GQDs [134]. These precursor molecules undergo controlled chemical reactions, such as pyrolysis, hydrothermal treatment, or solvothermal synthesis, leading to the formation of GQDs [133]. Both strategies offer unique advantages and can be tailored to achieve specific GQD properties, allowing for the customisation of size, shape, surface chemistry, and optical properties of the synthesised graphene quantum dots.

1.8 Graphene Properties

1.8.1 Electronic Properties

The exceptional electronic properties exhibited by graphene crystallites have been the main driving force behind the extensive attention given to graphene in both experimental and theoretical research. Graphene stands apart from conventional metals and semiconductors due to its distinctive properties arising from the electrons' ability to travel over long distances without being significantly hindered by electron-electron interactions or disorder. The extended mean free paths of electrons in graphene contribute to its unique physical structures and electronic properties. In the case of graphene, the Brillouin Zone is crucial in understanding its electronic properties. Due to the honeycomb lattice, graphene exhibits a hexagonal Brillouin Zone, which plays a fundamental role in determining the energy band structure and electronic behaviour of graphene. The combination of the honeycomb lattice structure and the hexagonal Brillouin Zone contributes to the exceptional electronic properties of graphene, including its distinctive band structure, high electron mobility, and the presence of Dirac cones at the corners of the Brillouin Zone, leading to massless Dirac fermions. In graphene, the charge carriers are effectively massless Dirac fermions, which results in a linear energy-momentum relationship. When a gate voltage is applied to graphene, the Fermi level can be shifted, enabling control over the carrier concentration and type. By modulating the gate voltage, graphene can switch between n-type and p-type behaviour, allowing for ambipolar conduction. This property is highly advantageous in electronic devices, as it offers flexibility in designing and controlling the functionality of graphene-based transistors and other electronic components. Another factor that is crucial for understanding the electronic properties of graphene e is the Quantum Hall effect (QHE). QHE is an intriguing electronic property exhibited by graphene [135]. QHE in graphene exhibits a distinct advantage over other 2D materials in terms of its temperature range [88]. Compared to other materials, the QHE in graphene can be observed at significantly higher temperatures [136]. The temperature range for the QHE in graphene is approximately ten times broader, allowing for the potential observation of this effect even at room temperature [137]. The Quantum Hall effect in graphene arises due to the unique electronic structure and the presence of massless Dirac fermions [138]. The Landau levels, which are discrete energy levels for charged particles in a magnetic field, exhibit a linear dispersion relation in graphene [138]. The quantised Hall resistance in graphene is characterised by plateaus at values that are integer multiples of the quantum resistance, i.e. h/e^2, where h is Planck's constant and e is the elementary charge [139]. This phenomenon provides a precise and accurate standard for resistance metrology and has significant implications for the study of fundamental physics, as well as potential applications in quantum devices and metrology [140].

1.8.2 Physical and Thermal Properties

Generally, the mechanical attributes of any crystalline solid are primarily determined by the inherent characteristics of its pristine crystal lattice. The structural defects within the solid, such as dislocations and grain boundaries, also play a significant role in influencing its mechanical behaviour. These defects can have a substantial impact on the material's strength, ductility, and other mechanical properties. The response of non-linear elasticity of graphene under tensile load can be described by the equation: $\sigma = E\varepsilon + D\varepsilon^2$, where σ represents the applied stress, ε denotes the uniaxial elastic strain, E represents the elastic modulus, and D accounts for elastic stiffness of third order [141]. Graphene exhibited Young's modulus of approximately 1.0 ± 0.1 TPa, indicating its remarkable stiffness. Furthermore, its breaking strength is estimated to be around 40 newtons per metre (N/m), showcasing its exceptional resistance to failure under applied stress [141]. Observations have demonstrated that graphene surpasses both graphite and carbon nanotubes in enhancing the Young's modulus and tensile strength of nanocomposites [142]. Graphene's superiority can be attributed to its two-dimensional geometry, high surface-to-volume ratio, and the resulting stronger interface interaction [143]. The unique two-dimensional structure of graphene allows for efficient load transfer and stress distribution within materials, leading to enhanced mechanical properties [22]. Additionally, at room temperature, graphene exhibits a thermal conductivity of approximately 5300 W/m*K. This high thermal conductivity makes graphene an excellent conductor of heat, enabling efficient thermal transport in various applications [144]. Graphene exhibits a specific surface area of up to 2630 (m^2/g) [145]. This exceptional surface area property enables its utilisation in microdetectors designed for gas molecules. By monitoring the significant change in resistance resulting from the adsorption and desorption of gas molecules, graphene-based microdetectors can accurately detect and analyse gas interactions at the molecular level [146].

1.9 Applications of Graphene

The exceptional characteristics of graphene have vastly broadened its potential uses in a wide range of industries, including electronics, medicine, composites, energy, and purification.

1.9.1 Applications in Electronics

Graphene's outstanding physical properties, such as its high electrical conductivity, flexibility, and transparency, make it an extremely attractive material for applications in the electronics industry. Graphene's high carrier mobility and exceptional electrical conductivity make it a promising candidate for the development of next-generation transistors and integrated circuits. Its ability to operate at high frequencies and its potential for ultra-miniaturisation could revolutionise electronics manufacturing [147]. Besides this, graphene's flexibility enables its integration into flexible and transparent electronic devices. These devices include flexible displays, wearable electronics, and bendable sensors, offering opportunities for innovative designs and improved user experiences [148].

Graphene-based sensors have shown great potential in various applications, including gas and chemical sensing, biosensing, and environmental monitoring [149, 150]. Graphene's high sensitivity, large surface area, and excellent electrical properties enable the development of highly sensitive and selective sensors with fast response times [150]. Also, graphene's optical properties, such as its high transparency and broadband absorption, make it promising for optoelectronic devices [151]. It can be used in photodetectors, light-emitting diodes (LEDs), solar cells, and optical modulators, enabling advancements in telecommunications, imaging, and energy harvesting [152–154].

1.9.2 Applications in Energy Storage

Since its discovery, graphene has emerged as a highly promising contender for energy storage materials, particularly in semiconductor applications as electrode materials. On complete use of its surface area, graphene exhibits a theoretical capacitance threshold of 550 Fg^{-1}, with an inherent capacitance of 21 μFcm^{-2} [155]. It is thus evident that graphene enables facile ion and electron accumulation and transportation, addressing the critical requirements for effective supercapacitor electrode materials. The high surface area and excellent electrical conductivity of graphene facilitate efficient ion adsorption and desorption at the electrode-electrolyte interface, leading to enhanced charge storage capabilities. Additionally, graphene's unique properties enable rapid ion and electron mobility, enabling fast charge and discharge rates in SC systems. These factors contribute to the promising application of graphene in electric double-layer capacitors (EDLCs), offering opportunities for the development of high-performance energy storage devices [156].

Interestingly, the exceptional properties of graphene, including high carrier mobility, low resistivity, excellent transmittance, and its compactly packed 2D structure, have made graphene-based materials highly attractive for utilisation in photovoltaic (PV) devices, potentially replacing conventional materials. Graphene shows promise as a viable alternative to commercial indium tin oxide (ITO) as a conductive electrode, particularly for flexible solar cells [157]. Graphene-based materials have the potential to serve as charge selective and transport components in buffer layers of solar cells [158]. These applications demonstrate the potential of graphene in enhancing the performance, stability, and flexibility of solar cell technologies.

1.9.3 Applications in Composite Materials

Graphene, with its exceptional properties and unique structure, holds immense potential for revolutionising the composite materials industry. Graphene has been widely used in composite in form of reinforcement to produce materials with enhanced properties. Its remarkable strength, stiffness, and lightweight nature make it an ideal candidate for enhancing the performance of composite materials. When incorporated into polymers, graphene can significantly improve their mechanical properties, including tensile strength and fracture toughness [159, 160]. The high aspect ratio of graphene allows for efficient stress transfer and load distribution within the composite, resulting in enhanced structural integrity and improved resistance to deformation and cracking [161].

1.9.4 Applications in Biomedical Applications

Graphene has tremendous potential in the field of medication delivery systems development. The delivery of therapeutic chemicals to specific target areas in the body can be facilitated through the use of graphene-based nanocarriers [162]. Due to its high surface area and biocompatibility, graphene exhibits a high drug-loading capacity and low toxicity [163]. Additionally, site-specific medication delivery is made possible by the ability to functionalise graphene with particular targeted moieties, increasing the effectiveness and efficiency of treatments [164]. Further, graphene can permit the development of stimuli-responsive drug delivery systems for precise control over therapeutic delivery by triggering the release of medicines at the intended area in response to external stimuli [165].

Furthermore, in tissue engineering, graphene has shown promise as a scaffold material for promoting cell growth and tissue regeneration. It is one of the most desirable 2D materials for making bioactive scaffolds due to its biocompatibility and capacity to promote cellular adhesion and proliferation [166]. The mechanical strength of graphene supports the structural integrity of modified tissues, while its high electrical behaviour can enhance the operation of electrically excitable tissues like cardiac and brain tissues [167, 168].

References

1. R. Peierls, Quelques propriétés typiques des corps solides. Annales de l'institut Henri Poincaré **5**(3), 177–222
2. L.D. Landau, On the theory of phase transitions. I. Zh. Eksp. Teor. Fiz. **11**, 19 (1937)
3. H.W. Kroto, J.R. Heath, S.C. O'Brien, R.F. Curl, R.E. Smalley, C60: Buckminsterfullerene. Nature **318**(6042), 162–163 (1985). https://doi.org/10.1038/318162a0
4. S. Iijima, Helical microtubules of graphitic carbon. Nature **354**(6348), 56–58 (1991). https://doi.org/10.1038/354056a0
5. A. Gupta, T. Sakthivel, S. Seal, Recent development in 2D materials beyond graphene. Prog. Mater. Sci. **73**, 44–126 (2015). https://doi.org/10.1016/j.pmatsci.2015.02.002
6. R.J.P. Williams, J.S. Rowlinson, A. Chapman (eds.), *Chemistry at Oxford* (Royal Society of Chemistry, Cambridge, 2008). https://doi.org/10.1039/9781847558855
7. D.A.C. Brownson, C.E. Banks, The electrochemistry of CVD graphene: Progress and prospects. Phys. Chem. Chem. Phys. **14**(23), 8264 (2012). https://doi.org/10.1039/c2cp40225d
8. S. Park, R.S. Ruoff, Chemical methods for the production of graphenes. Nat Nanotechnol **4**(4), 217–224 (2009). https://doi.org/10.1038/nnano.2009.58
9. D.R. Dreyer, R.S. Ruoff, C.W. Bielawski, From conception to realization: An Historical account of graphene and some perspectives for its future. Angew. Chem. Int. Ed. **49**(49), 9336–9344 (2010). https://doi.org/10.1002/anie.201003024
10. O.V. Prezhdo, Graphene — The ultimate surface material. Surf. Sci. **605**(17–18), 1607–1610 (2011). https://doi.org/10.1016/j.susc.2011.05.009
11. K.S. Novoselov, A.K. Geim, S.V. Morozov, D. Jiang, Y. Zhang, S.V. Dubonos, I.V. Grigorieva, A.A. Firsov, Electric field effect in atomically thin carbon films. Science (1979) **306**(5696), 666–669 (2004). https://doi.org/10.1126/science.1102896
12. K.S. Novoselov, D. Jiang, F. Schedin, T.J. Booth, V.V. Khotkevich, S.V. Morozov, A.K. Geim, Two-dimensional atomic crystals. Proc. Natl. Acad. Sci. **102**(30), 10451–10453 (2005). https://doi.org/10.1073/pnas.0502848102
13. K.S. Novoselov, Graphene: Materials in the flatland (Nobel lecture). Angew. Chem. Int. Ed. **50**(31), 6986–7002 (2011). https://doi.org/10.1002/anie.201101502
14. A.K. Geim, Random walk to graphene (Nobel lecture). Angew. Chem. Int. Ed. **50**(31), 6966–6985 (2011). https://doi.org/10.1002/anie.201101174
15. A. Hirsch, The era of carbon allotropes. Nat. Mater. **9**(11), 868–871 (2010). https://doi.org/10.1038/nmat2885
16. A.S. Lemine, M.M. Zagho, T.M. Altahtamouni, N. Bensalah, Graphene a promising electrode material for supercapacitors-A review. Int. J. Energy Res. **42**(14), 4284–4300 (2018). https://doi.org/10.1002/er.4170
17. M.I. Katsnelson, A. Fasolino, Graphene: Basic properties, in *2D Materials*, ed. by P. Avouris, T.F. Heinz, T. Low, (Cambridge University Press, 2017), pp. 7–24. https://doi.org/10.1017/9781316681619.002
18. A. Armano, S. Agnello, Two-dimensional carbon: A review of synthesis methods, and electronic, optical, and vibrational properties of single-layer graphene. C **5**(4), 67 (2019). https://doi.org/10.3390/c5040067
19. W. Zhang, C. Chai, Q. Fan, Y. Song, Y. Yang, A novel two-dimensional sp-sp2-sp3 hybridized carbon nanostructure with a negative in-plane Poisson ratio and high electron mobility. Comput. Mater. Sci. **185**, 109904 (2020). https://doi.org/10.1016/j.commatsci.2020.109904
20. A. Milani, A.L. Bassi, V. Russo, M. Tommasini, C.S. Casari, Linear carbon: From 1D Carbyne to 2D hybrid sp-sp 2 nanostructures beyond graphene, in *Handbook of Graphene*, (Wiley, 2019), pp. 297–340. https://doi.org/10.1002/9781119468455.ch48
21. S. Jung, P. C. Modelling, E. Devices, É. D. N, M. Compacte, and À. B. Physique, "Physically-based compact modelling of organic electronic devices," 2016
22. Z. Zhen, H. Zhu, Structure and properties of graphene, in *Graphene*, (Elsevier, 2018), pp. 1–12. https://doi.org/10.1016/B978-0-12-812651-6.00001-X

23. A.V. Klekachev, A. Nourbakhsh, I. Asselberghs, A.L. Stesmans, M.M. Heyns, S. De Gendt, Graphene transistors and photodetectors. Electrochem Soc Interface. **22**(1), 63–68 (2013). https://doi.org/10.1149/2.F07131if
24. D. Grasseschi, W.C. Silva, R. de Souza Paiva, L.D. Starke, A.S. do Nascimento, Surface coordination chemistry of graphene: Understanding the coordination of single transition metal atoms. Coord. Chem. Rev. **422**, 213469 (2020). https://doi.org/10.1016/j.ccr.2020.213469
25. T.M. Magne, T. de Oliveira Vieira, L.M.R. Alencar, F.F.M. Junior, S. Gemini-Piperni, S.V. Carneiro, L.M.U.D. Fechine, R.M. Freire, K. Golokhvast, P. Metrangolo, P.B.A. Fechine, R. Santos-Oliveira, Graphene and its derivatives: Understanding the main chemical and medicinal chemistry roles for biomedical applications. J. Nanostructure Chem. (2021). https://doi.org/10.1007/s40097-021-00444-3
26. G. Yang, L. Li, W.B. Lee, M.C. Ng, Structure of graphene and its disorders: A review. Sci. Technol. Adv. Mater. **19**(1), 613–648 (2018). https://doi.org/10.1080/14686996.2018.1494493
27. M. Tahriri, M. Del Monico, A. Moghanian, M. Tavakkoli Yaraki, R. Torres, A. Yadegari, L. Tayebi, Graphene and its derivatives: Opportunities and challenges in dentistry. Mater. Sci. Eng. C **102**, 171–185 (2019). https://doi.org/10.1016/j.msec.2019.04.051
28. M. Li, C. Wang, Preparation and characterization of GO/PEG photo-thermal conversion form-stable composite phase change materials. Renew. Energy **141**, 1005–1012 (2019). https://doi.org/10.1016/j.renene.2019.03.141
29. V. Karthik, P. Selvakumar, P. Senthil Kumar, D.-V.N. Vo, M. Gokulakrishnan, P. Keerthana, V. Tamil Elakkiya, R. Rajeswari, Graphene-based materials for environmental applications: A review. Environ. Chem. Lett. **19**(5), 3631–3644 (2021). https://doi.org/10.1007/s10311-021-01262-3
30. N. Kumar, R. Salehiyan, V. Chauke, O. Joseph Botlhoko, K. Setshedi, M. Scriba, M. Masukume, S. Sinha Ray, Top-down synthesis of graphene: A comprehensive review. FlatChem **27**, 100224 (2021). https://doi.org/10.1016/j.flatc.2021.100224
31. E. Gao, S.-Z. Lin, Z. Qin, M.J. Buehler, X.-Q. Feng, Z. Xu, Mechanical exfoliation of two-dimensional materials. J. Mech. Phys. Solids **115**, 248–262 (2018). https://doi.org/10.1016/j.jmps.2018.03.014
32. H. Hiura, T.W. Ebbesen, J. Fujita, K. Tanigaki, T. Takada, Role of sp3 defect structures in graphite and carbon nanotubes. Nature **367**(6459), 148–151 (1994). https://doi.org/10.1038/367148a0
33. T.W. Ebbesen, H. Hiura, Graphene in 3-dimensions: Towards graphite origami. Adv. Mater. **7**(6), 582–586 (1995). https://doi.org/10.1002/adma.19950070618
34. T.M. Bernhardt, B. Kaiser, K. Rademann, Formation of superperiodic patterns on highly oriented pyrolytic graphite by manipulation of nanosized graphite sheets with the STM tip. Surf. Sci. **408**(1–3), 86–94 (1998). https://doi.org/10.1016/S0039-6028(98)00152-6
35. X. Lu, M. Yu, H. Huang, R.S. Ruoff, Tailoring graphite with the goal of achieving single sheets. Nanotechnology **10**(3), 269–272 (1999). https://doi.org/10.1088/0957-4484/10/3/308
36. L. Ci, L. Song, D. Jariwala, A.L. ElÃas, W. Gao, M. Terrones, P.M. Ajayan, Graphene shape control by multistage cutting and transfer. Adv. Mater. **21**(44), 4487–4491 (2009). https://doi.org/10.1002/adma.200900942
37. X. Liang, A.S.P. Chang, Y. Zhang, B.D. Harteneck, H. Choo, D.L. Olynick, S. Cabrini, Electrostatic force assisted exfoliation of prepatterned few-layer graphenes into device sites. Nano Lett. **9**(1), 467–472 (2009). https://doi.org/10.1021/nl803512z
38. X. Liang, Z. Fu, S.Y. Chou, Graphene transistors fabricated via transfer-printing in device active-areas on large wafer. Nano Lett. **7**(12), 3840–3844 (2007). https://doi.org/10.1021/nl072566s
39. J.-H. Chen, M. Ishigami, C. Jang, D.R. Hines, M.S. Fuhrer, E.D. Williams, Printed graphene circuits. Adv. Mater. **19**(21), 3623–3627 (2007). https://doi.org/10.1002/adma.200701059
40. V. Huc, N. Bendiab, N. Rosman, T. Ebbesen, C. Delacour, V. Bouchiat, Large and flat graphene flakes produced by epoxy bonding and reverse exfoliation of highly oriented pyrolytic graphite. Nanotechnology **19**(45), 455601 (2008). https://doi.org/10.1088/0957-4484/19/45/455601

41. B. Jayasena, S. Subbiah, A novel mechanical cleavage method for synthesizing few-layer graphenes. Nanoscale Res. Lett. **6**(1), 95 (2011). https://doi.org/10.1186/1556-276X-6-95
42. J. Chen, M. Duan, G. Chen, Continuous mechanical exfoliation of graphene sheets via three-roll mill. J. Mater. Chem. **22**(37), 19625 (2012). https://doi.org/10.1039/c2jm33740a
43. M. Yi, Z. Shen, A review on mechanical exfoliation for the scalable production of graphene. J Mater Chem A Mater **3**(22), 11700–11715 (2015). https://doi.org/10.1039/C5TA00252D
44. K.R. Paton, E. Varrla, C. Backes, R.J. Smith, U. Khan, A. O'Neill, C. Boland, M. Lotya, O.M. Istrate, P. King, T. Higgins, S. Barwich, P. May, P. Puczkarski, I. Ahmed, M. Moebius, H. Pettersson, E. Long, J. Coelho, S.E. O'Brien, E.K. McGuire, B.M. Sanchez, G.S. Duesberg, N. McEvoy, T.J. Pennycook, C. Downing, A. Crossley, V. Nicolosi, J.N. Coleman, Scalable production of large quantities of defect-free few-layer graphene by shear exfoliation in liquids. Nat. Mater. **13**(6), 624–630 (2014). https://doi.org/10.1038/nmat3944
45. F. Rouzafzay, R. Shidpour, M.Z.M. Al-Abri, F. Qaderi, A. Ahmadi, M.T.Z. Myint, Graphene@ZnO nanocompound for short-time water treatment under sun-simulated irradiation: Effect of shear exfoliation of graphene using kitchen blender on photocatalytic degradation. J. Alloys Compd. **829**, 154614 (2020). https://doi.org/10.1016/j.jallcom.2020.154614
46. A. Niftaliyeva, E. Pehlivan, S. Polat, A. Avci, Chemical synthesis of single-layer graphene by using ball milling compared with NaBH 4 and hydroquinone reductants. Micro Nano Lett **13**(10), 1412–1416 (2018). https://doi.org/10.1049/mnl.2018.5165
47. Y. Arao, R. Kuwahara, K. Ohno, J. Tanks, K. Aida, M. Kubouchi, S. Takeda, Mass production of low-boiling point solvent- and water-soluble graphene by simple salt-assisted ball milling. Nanoscale Adv **1**(12), 4955–4964 (2019). https://doi.org/10.1039/C9NA00463G
48. S.S. Shams, R. Zhang, J. Zhu, Graphene synthesis: A review. Mater. Sci.-Pol. **33**(3), 566–578 (2015). https://doi.org/10.1515/msp-2015-0079
49. L. Liu, Z. Xiong, D. Hu, G. Wu, P. Chen, Production of high quality single- or few-layered graphene by solid exfoliation of graphite in the presence of ammonia borane. Chem. Commun. **49**(72), 7890 (2013). https://doi.org/10.1039/c3cc43670e
50. V. León, M. Quintana, M.A. Herrero, J.L.G. Fierro, A. de la Hoz, M. Prato, E. Vázquez, Few-layer graphenes from ball-milling of graphite with melamine. Chem. Commun. **47**(39), 10936 (2011). https://doi.org/10.1039/c1cc14595a
51. G. Yoon, D.-H. Seo, K. Ku, J. Kim, S. Jeon, K. Kang, Factors affecting the exfoliation of graphite intercalation compounds for graphene synthesis. Chem. Mater. **27**(6), 2067–2073 (2015). https://doi.org/10.1021/cm504511b
52. J.-C. An, E.J. Lee, I. Hong, Preparation of the spheroidized graphite-derived multi-layered graphene via GIC (graphite intercalation compound) method. J. Ind. Eng. Chem. **47**, 56–61 (2017). https://doi.org/10.1016/j.jiec.2016.12.017
53. Y.H. Bae, T.S. Kwon, M.J. Yu, B.C. Lee, S.R. Kim, Acoustic characteristics and thermal properties of polycarbonate/(graphite intercalation compound) composites. Polymer (Korea) **41**(2), 189–195 (2017). https://doi.org/10.7317/PK.2017.41.2.189
54. V. Singh, D. Joung, L. Zhai, S. Das, S.I. Khondaker, S. Seal, Graphene based materials: Past, present and future. Prog. Mater. Sci. **56**(8), 1178–1271 (2011). https://doi.org/10.1016/j.pmatsci.2011.03.003
55. J.C. Chacón-Torres, L. Wirtz, T. Pichler, Raman spectroscopy of graphite intercalation compounds: Charge transfer, strain, and electron-phonon coupling in graphene layers. Phys. Status Solidi B **251**(12), 2337–2355 (2014). https://doi.org/10.1002/pssb.201451477
56. K. Yan, L. Fu, H. Peng, Z. Liu, Designed CVD growth of graphene via process engineering. Acc. Chem. Res. **46**(10), 2263–2274 (2013). https://doi.org/10.1021/ar400057n
57. M.H. Ani, M.A. Kamarudin, A.H. Ramlan, E. Ismail, M.S. Sirat, M.A. Mohamed, M.A. Azam, A critical review on the contributions of chemical and physical factors toward the nucleation and growth of large-area graphene. J. Mater. Sci. **53**(10), 7095–7111 (2018). https://doi.org/10.1007/s10853-018-1994-0
58. C.R.S.V. Boas, B. Focassio, E. Marinho, D.G. Larrude, M.C. Salvadori, C.R. Leão, D.J. dos Santos, Characterization of nitrogen doped graphene bilayers synthesized by fast, low tem-

perature microwave plasma-enhanced chemical vapour deposition. Sci. Rep. **9**(1), 13715 (2019). https://doi.org/10.1038/s41598-019-49900-9
59. Y. Dong, S. Guo, H. Mao, C. Xu, Y. Xie, C. Cheng, X. Mao, J. Deng, G. Pan, J. Sun, The growth of graphene on Ni–Cu alloy thin films at a low temperature and its carbon diffusion mechanism. Nano **9**(11), 1633 (2019). https://doi.org/10.3390/nano9111633
60. T. Akimoto and K. Ueno, High crystallinity multilayer graphene deposited by a low-temperature CVD using Ni catalyst with applying current. 2019 Electron Devices Technology and Manufacturing Conference (EDTM), Mar. 2019, pp. 351–353. doi: https://doi.org/10.1109/EDTM.2019.8731128
61. P. Sutter, How silicon leaves the scene. Nat. Mater. **8**(3), 171–172 (2009). https://doi.org/10.1038/nmat2392
62. T. Ohta, A. Bostwick, J.L. McChesney, T. Seyller, K. Horn, E. Rotenberg, Interlayer interaction and electronic screening in multilayer graphene investigated with angle-resolved photoemission spectroscopy. Phys. Rev. Lett. **98**(20), 206802 (2007). https://doi.org/10.1103/PhysRevLett.98.206802
63. M.S.A. Bhuyan, M.N. Uddin, M.M. Islam, F.A. Bipasha, S.S. Hossain, Synthesis of graphene. Int. Nano Lett. **6**(2), 65–83 (2016). https://doi.org/10.1007/s40089-015-0176-1
64. S. Amini, J. Garay, G. Liu, A.A. Balandin, R. Abbaschian, Growth of large-area graphene films from metal-carbon melts. J. Appl. Phys. **108**(9), 094321 (2010). https://doi.org/10.1063/1.3498815
65. P.W. Sutter, J.-I. Flege, E.A. Sutter, Epitaxial graphene on ruthenium. Nat. Mater. **7**(5), 406–411 (2008). https://doi.org/10.1038/nmat2166
66. I. Pletikosić, M. Kralj, P. Pervan, R. Brako, J. Coraux, A.T. N'Diaye, C. Busse, T. Michely, Dirac cones and minigaps for graphene on Ir(111). Phys. Rev. Lett. **102**(5), 056808 (2009). https://doi.org/10.1103/PhysRevLett.102.056808
67. K.S. Kim, Y. Zhao, H. Jang, S.Y. Lee, J.M. Kim, K.S. Kim, J.H. Ahn, P. Kim, J.Y. Choi, B.H. Hong, Large-scale pattern growth of graphene films for stretchable transparent electrodes. Nature **457**(7230), 706–710 (2009). https://doi.org/10.1038/nature07719
68. Z. Fogarassy, M.H. Rümmeli, S. Gorantla, A. Bachmatiuk, G. Dobrik, K. Kamarás, L.P. Biró, K. Havancsák, J.L. Lábár, Dominantly epitaxial growth of graphene on Ni (1 1 1) substrate. Appl. Surf. Sci. **314**, 490–499 (2014). https://doi.org/10.1016/j.apsusc.2014.06.197
69. A. Chakrabarti, J. Lu, J.C. Skrabutenas, T. Xu, Z. Xiao, J.A. Maguire, N.S. Hosmane, Conversion of carbon dioxide to few-layer graphene. J. Mater. Chem. **21**(26), 9491 (2011). https://doi.org/10.1039/c1jm11227a
70. E. Cuadros-Lugo, M. Piñon-Espitia, H.A. Martinez-Rodríguez, D. Lardizabal-Gutierrez, I. Estrada-Guel, J.M. Herrera-Ramirez, C. Carreño-Gallardo, Turbostratic carbon/graphene prepared via the dry ice in flames method and its purification using different routes: A comparative study. Materials **15**(7), 2501 (2022). https://doi.org/10.3390/ma15072501
71. J. Zhang, T. Tian, Y. Chen, Y. Niu, J. Tang, L.-C. Qin, Synthesis of graphene from dry ice in flames and its application in supercapacitors. Chem. Phys. Lett. **591**, 78–81 (2014). https://doi.org/10.1016/j.cplett.2013.11.014
72. K. Chu, F. Wang, Y.-B. Li, X.-H. Wang, D.-J. Huang, Z.-R. Geng, Interface and mechanical/thermal properties of graphene/copper composite with Mo_2C nanoparticles grown on graphene. Compos. Part A Appl. Sci. Manuf. **109**(March), 267–279 (2018). https://doi.org/10.1016/j.compositesa.2018.03.014
73. C. Salvo, R.V. Mangalaraja, R. Udayabashkar, M. Lopez, C. Aguilar, Enhanced mechanical and electrical properties of novel graphene reinforced copper matrix composites. J. Alloys Compd. **777**, 309–316 (2019). https://doi.org/10.1016/j.jallcom.2018.10.357
74. F. Chen, J. Ying, Y. Wang, S. Du, Z. Liu, Q. Huang, Effects of graphene content on the microstructure and properties of copper matrix composites. Carbon N Y **96**, 836–842 (2016). https://doi.org/10.1016/j.carbon.2015.10.023
75. M. Yang, L. Weng, H. Zhu, T. Fan, D. Zhang, Simultaneously enhancing the strength, ductility and conductivity of copper matrix composites with graphene nanoribbons. Carbon N Y **118**, 250–260 (2017). https://doi.org/10.1016/j.carbon.2017.03.055

76. H. Jang, S. Yoo, M. Quevedo, H. Choi, Effect of processing route on mechanical and thermal properties of few-layered graphene (FLG)-reinforced copper matrix composites. J. Alloys Compd. **754**, 7–13 (2018). https://doi.org/10.1016/j.jallcom.2018.04.272
77. D.L. Duong, G.H. Han, S.M. Lee, F. Gunes, E.S. Kim, S.T. Kim, H. Kim, Q.H. Ta, K.P. So, S.J. Yoon, S.J. Chae, Y.W. Jo, M.H. Park, S.H. Chae, S.C. Lim, J.Y. Choi, Y.H. Lee, Probing graphene grain boundaries with optical microscopy. Nature **490**(7419), 235–241 (2012)
78. J. Kim, K.E. Lee, K.H. Kim, S. Wi, S. Lee, S. Nam, C. Kim, S.O. Kim, B. Park, Single-layer graphene-wrapped Li4Ti5O12 anode with superior lithium storage capability. Carbon N Y **114**, 275–283 (2017). https://doi.org/10.1016/j.carbon.2016.12.022
79. M. Khenfouch, U. Buttner, M. Baïtoul, M. Maaza, Synthesis and characterization of mass produced high quality few layered graphene sheets via a chemical method. Graphene **03**(02), 7–13 (2014). https://doi.org/10.4236/graphene.2014.32002
80. X. Gao, H. Yue, E. Guo, S. Zhang, L. Yao, X. Lin, B. Wang, E. Guan, Tribological properties of copper matrix composites reinforced with homogeneously dispersed graphene nanosheets. J. Mater. Sci. Technol. **34**(10), 1925–1931 (2018). https://doi.org/10.1016/j.jmst.2018.02.010
81. D. Qu, F.Z. Li, H. Bin Zhang, Q. Wang, T.L. Zhou, C.F. Hu, R.J. Xie, Preparation of graphene nanosheets/copper composite by spark plasma sintering. Adv. Mater. Res. **833**, 276–279 (2013). https://doi.org/10.4028/www.scientific.net/AMR.833.276
82. X. Si, M. Li, F. Chen, P. Eklund, J. Xue, F. Huang, S. Du, Q. Huang, Effect of carbide interlayers on the microstructure and properties of graphene-nanoplatelet-reinforced copper matrix composites. Mater. Sci. Eng. A **708**(August), 311–318 (2017). https://doi.org/10.1016/j.msea.2017.10.015
83. C.F. Escalante, C. Fabian, E. Sierra, Fundamentals of transmission electron microscopy, the technique with the best resolution in the world. Screen (London) **no. 9**, 10 (2019)
84. M. Ziatdinov, O. Dyck, A. Maksov, X. Li, X. Sang, K. Xiao, R.R. Unocic, R. Vasudevan, S. Jesse, S.V. Kalinin, Deep learning of atomically resolved scanning transmission electron microscopy images: Chemical identification and tracking local transformations. ACS Nano **11**(12), 12742–12752 (2017). https://doi.org/10.1021/acsnano.7b07504
85. F.T. Johra, J.-W. Lee, W.-G. Jung, Facile and safe graphene preparation on solution based platform. J. Ind. Eng. Chem. **20**(5), 2883–2887 (2014). https://doi.org/10.1016/j.jiec.2013.11.022
86. D.-M. Chen, P.M. Shenai, Y. Zhao, Tight binding description on the band gap opening of pyrene-dispersed graphene. Phys. Chem. Chem. Phys. **13**(4), 1515–1520 (2011). https://doi.org/10.1039/C0CP00909A
87. A.C. Ferrari, J.C. Meyer, V. Scardaci, C. Casiraghi, M. Lazzeri, F. Mauri, S. Piscanec, D. Jiang, K.S. Novoselov, S. Roth, A.K. Geim, Raman Spectrum of graphene and graphene layers. Phys. Rev. Lett. **97**(18), 187401 (2006). https://doi.org/10.1103/PhysRevLett.97.187401
88. Z. Jiang, Y. Zhang, Y.-W. Tan, H.L. Stormer, P. Kim, Quantum hall effect in graphene. Solid State Commun. **143**(1–2), 14–19 (2007). https://doi.org/10.1016/j.ssc.2007.02.046
89. M.O. Goerbig, Electronic properties of graphene in a strong magnetic field. Rev. Mod. Phys. **83**(4), 1193–1243 (2011). https://doi.org/10.1103/RevModPhys.83.1193
90. Y. Yang, W. Li, E. Kroner, E. Arzt, B. Bhushan, L. Benameur, L. Wei, A. Botta, Y. Lu, J. Lou, D. Jena, M. Nosonovsky, B. Bhushan, T. Søndergaard, P.K. Sekhar, S. Bhansali, A.A. Trusov, Graphene, in *Encyclopedia of Nanotechnology*, (Springer Netherlands, Dordrecht, 2012), pp. 968–978. https://doi.org/10.1007/978-90-481-9751-4_373
91. A. Das, S. Pisana, B. Chakraborty, S. Piscanec, S.K. Saha, U.V. Waghmare, K.S. Novoselov, H.R. Krishnamurthy, A.K. Geim, A.C. Ferrari, A.K. Sood, Monitoring dopants by Raman scattering in an electrochemically top-gated graphene transistor. Nat. Nanotechnol. **3**(4), 210–215 (2008). https://doi.org/10.1038/nnano.2008.67
92. W. Fang, A.L. Hsu, Y. Song, J. Kong, A review of large-area bilayer graphene synthesis by chemical vapor deposition. Nanoscale **7**(48), 20335–20351 (2015). https://doi.org/10.1039/C5NR04756K
93. E. McCann, M. Koshino, The electronic properties of bilayer graphene. Rep. Prog. Phys. **76**(5), 056503 (2013). https://doi.org/10.1088/0034-4885/76/5/056503

94. Y. Zhang, T.-T. Tang, C. Girit, Z. Hao, M.C. Martin, A. Zettl, M.F. Crommie, Y.R. Shen, F. Wang, Direct observation of a widely tunable bandgap in bilayer graphene. Nature **459**(7248), 820–823 (2009). https://doi.org/10.1038/nature08105
95. P. San-Jose, E. Prada, E. McCann, H. Schomerus, Pseudospin valve in bilayer graphene: Towards graphene-based pseudospintronics. Phys. Rev. Lett. **102**(24), 247204 (2009). https://doi.org/10.1103/PhysRevLett.102.247204
96. B. Sensale-Rodriguez, R. Yan, M.M. Kelly, T. Fang, K. Tahy, W.S. Hwang, D. Jena, L. Liu, H.G. Xing, Broadband graphene terahertz modulators enabled by intraband transitions. Nat. Commun. **3**(1), 780 (2012). https://doi.org/10.1038/ncomms1787
97. L. Zhou, L. Fox, M. Włodek, L. Islas, A. Slastanova, E. Robles, O. Bikondoa, R. Harniman, N. Fox, M. Cattelan, W.H. Briscoe, Surface structure of few layer graphene. Carbon N Y **136**, 255–261 (2018). https://doi.org/10.1016/j.carbon.2018.04.089
98. C. Zhang, G. Zeng, R. Zhang, Y. Tang, Q. Liu, T. Jiang, Tunable nonlinear optical responses of few-layer graphene through lithium intercalation. Nano **10**(10), 2661–2669 (2021). https://doi.org/10.1515/nanoph-2021-0173
99. K.R. Paton, J. Anderson, A.J. Pollard, T. Sainsbury, Production of few-layer graphene by microfluidization. Mater Res Express **4**(2), 025604 (2017). https://doi.org/10.1088/2053-1591/aa5b24
100. B.-H. Wee, T.-F. Wu, J.-D. Hong, Facile and scalable synthesis method for high-quality few-layer graphene through solution-based exfoliation of graphite. ACS Appl. Mater. Interfaces **9**(5), 4548–4557 (2017). https://doi.org/10.1021/acsami.6b11771
101. H.J. Park, J. Meyer, S. Roth, V. Skákalová, Growth and properties of few-layer graphene prepared by chemical vapor deposition. Carbon N Y **48**(4), 1088–1094 (2010). https://doi.org/10.1016/j.carbon.2009.11.030
102. D.S. Shin, H.G. Kim, H.S. Ahn, H.Y. Jeong, Y.-J. Kim, D. Odkhuu, N. Tsogbadrakh, H.-B.-R. Lee, B.H. Kim, Distribution of oxygen functional groups of graphene oxide obtained from low-temperature atomic layer deposition of titanium oxide. RSC Adv. **7**(23), 13979–13984 (2017). https://doi.org/10.1039/C7RA00114B
103. V. Kumar, A. Kumar, D.-J. Lee, S.-S. Park, Estimation of number of graphene layers using different methods: A focused review. Materials **14**(16), 4590 (2021). https://doi.org/10.3390/ma14164590
104. K.P. Loh, Q. Bao, G. Eda, M. Chhowalla, Graphene oxide as a chemically tunable platform for optical applications. Nat. Chem. **2**(12), 1015–1024 (2010). https://doi.org/10.1038/nchem.907
105. X. Chen, H. Guo, T. Wang, M. Lu, T. Wang, In-situ fabrication of reduced graphene oxide (rGO)/ZnO heterostructure: Surface functional groups induced electrical properties. Electrochim. Acta **196**, 558–564 (2016). https://doi.org/10.1016/j.electacta.2016.02.201
106. W. Du, H. Wu, H. Chen, G. Xu, C. Li, Graphene oxide in aqueous and nonaqueous media: Dispersion behaviour and solution chemistry. Carbon N Y **158**, 568–579 (2020). https://doi.org/10.1016/j.carbon.2019.11.027
107. S. Pei, H.-M. Cheng, The reduction of graphene oxide. Carbon N Y **50**(9), 3210–3228 (2012). https://doi.org/10.1016/j.carbon.2011.11.010
108. D.G. Papageorgiou, I.A. Kinloch, R.J. Young, Graphene/elastomer nanocomposites. Carbon N Y **95**, 460–484 (2015). https://doi.org/10.1016/j.carbon.2015.08.055
109. S. Park, J. An, J.R. Potts, A. Velamakanni, S. Murali, R.S. Ruoff, Hydrazine-reduction of graphite- and graphene oxide. Carbon N Y **49**(9), 3019–3023 (2011). https://doi.org/10.1016/j.carbon.2011.02.071
110. S.C. Ray, Application and uses of graphene oxide and reduced graphene oxide, in *Applications of Graphene and Graphene-Oxide Based Nanomaterials*, (Elsevier, 2015), pp. 39–55. https://doi.org/10.1016/B978-0-323-37521-4.00002-9
111. J. Wu, H. Lin, D.J. Moss, K.P. Loh, B. Jia, Graphene oxide for photonics, electronics and optoelectronics. Nat Rev Chem **7**(3), 162–183 (2023). https://doi.org/10.1038/s41570-022-00458-7

112. M. Ramezani, M. Alibolandi, M. Nejabat, F. Charbgoo, S.M. Taghdisi, K. Abnous, Graphene-based hybrid nanomaterials for biomedical applications, in *Biomedical Applications of Graphene and 2D Nanomaterials*, (Elsevier, 2019), pp. 119–141. https://doi.org/10.1016/B978-0-12-815889-0.00006-4
113. A. Hazra, S. Basu, Graphene nanoribbon as potential on-chip interconnect material—A review. C (Basel) **4**(3), 49 (2018). https://doi.org/10.3390/c4030049
114. J. Kumar, K. Peramaiya, N. Bernaurdshaw, J.T. David, Functionalization of zigzag graphene nanoribbon with DNA nucleobases-A DFT study. Appl. Surf. Sci. **496**, 143667 (2019). https://doi.org/10.1016/j.apsusc.2019.143667
115. H. Bu, Y. Chen, M. Zou, H. Yi, K. Bi, Z. Ni, Atomistic simulations of mechanical properties of graphene nanoribbons. Phys. Lett. A **373**(37), 3359–3362 (2009). https://doi.org/10.1016/j.physleta.2009.07.048
116. U. Rajaji, Graphene nanoribbons in electrochemical sensors and biosensors: A review. Int. J. Electrochem. Sci., 6643–6654 (2018). https://doi.org/10.20964/2018.07.51
117. A.P. Johnson, H.V. Gangadharappa, K. Pramod, Graphene nanoribbons: A promising nanomaterial for biomedical applications. J. Control. Release **325**, 141–162 (2020). https://doi.org/10.1016/j.jconrel.2020.06.034
118. A. Narita, X. Feng, Y. Hernandez, S.A. Jensen, M. Bonn, H. Yang, I.A. Verzhbitskiy, C. Casiraghi, M.R. Hansen, A.H.R. Koch, G. Fytas, O. Ivasenko, B. Li, K.S. Mali, T. Balandina, S. Mahesh, S. De Feyter, K. Müllen, Synthesis of structurally well-defined and liquid-phase-processable graphene nanoribbons. Nat. Chem. **6**(2), 126–132 (2014). https://doi.org/10.1038/nchem.1819
119. Y. Huang, Y. Mai, U. Beser, J. Teyssandier, G. Velpula, H. van Gorp, L.A. Straasø, M.R. Hansen, D. Rizzo, C. Casiraghi, R. Yang, G. Zhang, D. Wu, F. Zhang, D. Yan, S. De Feyter, K. Müllen, X. Feng, Poly(ethylene oxide) functionalized graphene nanoribbons with excellent solution processability. J. Am. Chem. Soc. **138**(32), 10136–10139 (2016). https://doi.org/10.1021/jacs.6b07061
120. Y. Huang, F. Xu, L. Ganzer, F.V.A. Camargo, T. Nagahara, J. Teyssandier, H. Van Gorp, K. Basse, L.A. Straasø, V. Nagyte, C. Casiraghi, M.R. Hansen, S. De Feyter, D. Yan, K. Müllen, X. Feng, G. Cerullo, Y. Mai, Intrinsic properties of single graphene nanoribbons in solution: Synthetic and spectroscopic studies. J. Am. Chem. Soc. **140**(33), 10416–10420 (2018). https://doi.org/10.1021/jacs.8b06028
121. M. Berahman, M. Asad, M. Sanaee, M.H. Sheikhi, Optical properties of chiral graphene nanoribbons: A first principle study. Opt. Quant. Electron. **47**(10), 3289–3300 (2015). https://doi.org/10.1007/s11082-015-0207-1
122. T.I. Mazilova, E.V. Sadanov, I.M. Mikhailovskij, Tensile strength of graphene nanoribbons: An experimental approach. Mater. Lett. **242**, 17–19 (2019). https://doi.org/10.1016/j.matlet.2019.01.051
123. N. Merino-Díez, A. Garcia-Lekue, E. Carbonell-Sanromà, J. Li, M. Corso, L. Colazzo, F. Sedona, D. Sánchez-Portal, J.I. Pascual, D.G. de Oteyza, Width-dependent band gap in armchair graphene nanoribbons reveals Fermi level pinning on Au(111). ACS Nano **11**(11), 11661–11668 (2017). https://doi.org/10.1021/acsnano.7b06765
124. Y.R. Kumar, K. Deshmukh, K.K. Sadasivuni, S.K.K. Pasha, Graphene quantum dot based materials for sensing, bio-imaging and energy storage applications: A review. RSC Adv. **10**(40), 23861–23898 (2020). https://doi.org/10.1039/D0RA03938A
125. Z. Jin, P. Owour, S. Lei, L. Ge, Graphene, graphene quantum dots and their applications in optoelectronics. Curr. Opin. Colloid Interface Sci. **20**(5–6), 439–453 (2015). https://doi.org/10.1016/j.cocis.2015.11.007
126. D. Pan, J. Zhang, Z. Li, M. Wu, Hydrothermal route for cutting graphene sheets into blue-luminescent graphene quantum dots. Adv. Mater. **22**(6), 734–738 (2010). https://doi.org/10.1002/adma.200902825
127. K.A. Ritter, J.W. Lyding, The influence of edge structure on the electronic properties of graphene quantum dots and nanoribbons. Nat. Mater. **8**(3), 235–242 (2009). https://doi.org/10.1038/nmat2378

128. L. Wang, Y. Wang, T. Xu, H. Liao, C. Yao, Y. Liu, Z. Li, Z. Chen, D. Pan, L. Sun, M. Wu, Gram-scale synthesis of single-crystalline graphene quantum dots with superior optical properties. Nat. Commun. **5**(1), 5357 (2014). https://doi.org/10.1038/ncomms6357
129. N. Nesakumar, S. Srinivasan, S. Alwarappan, Graphene quantum dots: Synthesis, properties, and applications to the development of optical and electrochemical sensors for chemical sensing. Microchim. Acta **189**(7), 258 (2022). https://doi.org/10.1007/s00604-022-05353-y
130. S.-G. Yim, Y. Kim, Y.-E. Kang, B. Moon, E. Jung, S. Yang, Size fractionation of fluorescent graphene quantum dots using a cross-flow membrane filtration system. Nano **8**(11), 959 (2018). https://doi.org/10.3390/nano8110959
131. R. Rabeya, S. Mahalingam, A. Manap, M. Satgunam, M. Akhtaruzzaman, C.H. Chia, Structural defects in graphene quantum dots: A review. Int. J. Quantum Chem. **122**(12) (2022). https://doi.org/10.1002/qua.26900
132. Y. Qin, Y. Cheng, L. Jiang, X. Jin, M. Li, X. Luo, G. Liao, T. Wei, Q. Li, Top-down strategy toward versatile graphene quantum dots for organic/inorganic hybrid solar cells. ACS Sustain. Chem. Eng. **3**(4), 637–644 (2015). https://doi.org/10.1021/sc500761n
133. P. Tian, L. Tang, K.S. Teng, S.P. Lau, Graphene quantum dots from chemistry to applications. Mater Today Chem **10**, 221–258 (2018). https://doi.org/10.1016/j.mtchem.2018.09.007
134. C. Zhao, X. Song, Y. Liu, Y. Fu, L. Ye, N. Wang, F. Wang, L. Li, M. Mohammadniaei, M. Zhang, Q. Zhang, J. Liu, Synthesis of graphene quantum dots and their applications in drug delivery. J Nanobiotechnol **18**(1), 142 (2020). https://doi.org/10.1186/s12951-020-00698-z
135. M.A. Hidalgo, Quantum hall effects in two-dimensional electron systems: A global approach. The European Physical Journal Plus **137**(1), 58 (2022). https://doi.org/10.1140/epjp/s13360-021-02173-6
136. A.M.R. Baker, J.A. Alexander-Webber, T. Altebaeumer, R.J. Nicholas, Energy relaxation for hot Dirac fermions in graphene and breakdown of the quantum hall effect. Phys. Rev. B **85**(11), 115403 (2012). https://doi.org/10.1103/PhysRevB.85.115403
137. M. Sang, J. Shin, K. Kim, K. Yu, Electronic and thermal properties of graphene and recent advances in graphene based electronics applications. Nano **9**(3), 374 (2019). https://doi.org/10.3390/nano9030374
138. T. Taychatanapat, K. Watanabe, T. Taniguchi, P. Jarillo-Herrero, Quantum hall effect and Landau-level crossing of Dirac fermions in trilayer graphene. Nat. Phys. **7**(8), 621–625 (2011). https://doi.org/10.1038/nphys2008
139. B. Özyilmaz, P. Jarillo-Herrero, D. Efetov, D.A. Abanin, L.S. Levitov, P. Kim, Electronic transport and quantum hall effect in bipolar graphene p–n–p junctions. Phys. Rev. Lett. **99**(16), 166804 (2007). https://doi.org/10.1103/PhysRevLett.99.166804
140. A.F. Rigosi, A.R. Panna, S.U. Payagala, M. Kruskopf, M.E. Kraft, G.R. Jones, B.-Y. Wu, H.-Y. Lee, Y. Yang, J. Hu, D.G. Jarrett, D.B. Newell, R.E. Elmquist, Graphene devices for tabletop and high-current quantized hall resistance standards. IEEE Trans. Instrum. Meas. **68**(6), 1870–1878 (2019). https://doi.org/10.1109/TIM.2018.2882958
141. C. Lee, X. Wei, J.W. Kysar, J. Hone, Measurement of the elastic properties and intrinsic strength of monolayer graphene. Science (1979) **321**(5887), 385–388 (2008). https://doi.org/10.1126/science.1157996
142. M.A. Rafiee, J. Rafiee, Z. Wang, H. Song, Z.-Z. Yu, N. Koratkar, Enhanced mechanical properties of nanocomposites at low graphene content. ACS Nano **3**(12), 3884–3890 (2009). https://doi.org/10.1021/nn9010472
143. V.B. Mbayachi, E. Ndayiragije, T. Sammani, S. Taj, E.R. Mbuta, A.U. Khan, Graphene synthesis, characterization and its applications: A review. Results Chem **3**, 100163 (2021). https://doi.org/10.1016/j.rechem.2021.100163
144. A.A. Balandin, S. Ghosh, W. Bao, I. Calizo, D. Teweldebrhan, F. Miao, C.N. Lau, Superior thermal conductivity of single-layer graphene. Nano Lett. **8**(3), 902–907 (2008). https://doi.org/10.1021/nl0731872
145. M.D. Stoller, S. Park, Y. Zhu, J. An, R.S. Ruoff, Graphene-based ultracapacitors. Nano Lett. **8**(10), 3498–3502 (2008). https://doi.org/10.1021/nl802558y

146. F. Schedin, A.K. Geim, S.V. Morozov, E.W. Hill, P. Blake, M.I. Katsnelson, K.S. Novoselov, Detection of individual gas molecules adsorbed on graphene. Nat. Mater. **6**(9), 652–655 (2007). https://doi.org/10.1038/nmat1967
147. P. Avouris, C. Dimitrakopoulos, Graphene: Synthesis and applications. Mater. Today **15**(3), 86–97 (2012). https://doi.org/10.1016/S1369-7021(12)70044-5
148. H. Kim, J.-H. Ahn, Graphene for flexible and wearable device applications. Carbon N Y **120**, 244–257 (2017). https://doi.org/10.1016/j.carbon.2017.05.041
149. A. Nag, A. Mitra, S.C. Mukhopadhyay, Graphene and its sensor-based applications: A review. Sens Actuators A Phys **270**, 177–194 (2018). https://doi.org/10.1016/j.sna.2017.12.028
150. C. Wang, Y. Wang, Z. Yang, N. Hu, Review of recent progress on graphene-based composite gas sensors. Ceram. Int. **47**(12), 16367–16384 (2021). https://doi.org/10.1016/j.ceramint.2021.02.144
151. F. Bonaccorso, Z. Sun, T. Hasan, A.C. Ferrari, Graphene photonics and optoelectronics. Nat. Photonics **4**(9), 611–622 (2010). https://doi.org/10.1038/nphoton.2010.186
152. A. Montanaro, W. Wei, D. De Fazio, U. Sassi, G. Soavi, P. Aversa, A.C. Ferrari, H. Happy, P. Legagneux, E. Pallecchi, Optoelectronic mixing with high-frequency graphene transistors. Nat. Commun. **12**(1), 2728 (2021). https://doi.org/10.1038/s41467-021-22943-1
153. X. Li, X. Zhang, H. Park, A. Di Bartolomeo, Editorial: Electronics and optoelectronics of graphene and related 2D materials. Front Mater **7** (2020). https://doi.org/10.3389/fmats.2020.00235
154. J. Wang, X. Mu, M. Sun, T. Mu, Optoelectronic properties and applications of graphene-based hybrid nanomaterials and van der Waals heterostructures. Appl. Mater. Today **16**, 1–20 (2019). https://doi.org/10.1016/j.apmt.2019.03.006
155. J. Xia, F. Chen, J. Li, N. Tao, Measurement of the quantum capacitance of graphene. Nat. Nanotechnol. **4**(8), 505–509 (2009). https://doi.org/10.1038/nnano.2009.177
156. H. Zhang, D. Yang, A. Lau, T. Ma, H. Lin, B. Jia, Hybridized graphene for supercapacitors: Beyond the limitation of pure graphene. Small **17**(12), 2007311 (2021). https://doi.org/10.1002/smll.202007311
157. Y. Woo, Transparent conductive electrodes based on graphene-related materials. Micromachines (Basel) **10**(1), 13 (2018). https://doi.org/10.3390/mi10010013
158. N.E. Safie, M.A. Azam, M.F.A. Aziz, M. Ismail, Recent progress of graphene-based materials for efficient charge transfer and device performance stability in perovskite solar cells. Int. J. Energy Res. **45**(2), 1347–1374 (2021). https://doi.org/10.1002/er.5876
159. J. Wang, F. Song, Y. Ding, M. Shao, The incorporation of graphene to enhance mechanical properties of polypropylene self-reinforced polymer composites. Mater. Des. **195**, 109073 (2020). https://doi.org/10.1016/j.matdes.2020.109073
160. A. Kumar, K. Sharma, A.R. Dixit, A review of the mechanical and thermal properties of graphene and its hybrid polymer nanocomposites for structural applications. J. Mater. Sci. **54**(8), 5992–6026 (2019). https://doi.org/10.1007/s10853-018-03244-3
161. E. Kandare, A.A. Khatibi, S. Yoo, R. Wang, J. Ma, P. Olivier, N. Gleizes, C.H. Wang, Improving the through-thickness thermal and electrical conductivity of carbon fibre/epoxy laminates by exploiting synergy between graphene and silver nano-inclusions. Compos. Part A Appl. Sci. Manuf. **69**, 72–82 (2015). https://doi.org/10.1016/j.compositesa.2014.10.024
162. A.D. Sontakke, S. Tiwari, M.K. Purkait, A comprehensive review on graphene oxide-based nanocarriers: Synthesis, functionalization and biomedical applications. FlatChem **38**, 100484 (2023). https://doi.org/10.1016/j.flatc.2023.100484
163. R. G. Bai and G. A. Husseini, Graphene-based drug delivery systems," in Biomimetic Nanoengineered Materials for Advanced Drug Delivery, Elsevier, 2019, pp. 149–168. doi: https://doi.org/10.1016/B978-0-12-814944-7.00011-4
164. A. Stergiou, R. Cantón-Vitoria, M.N. Psarrou, S.P. Economopoulos, N. Tagmatarchis, Functionalized graphene and targeted applications—Highlighting the road from chemistry to applications. Prog. Mater. Sci. **114**, 100683 (2020). https://doi.org/10.1016/j.pmatsci.2020.100683

165. P. Vinchhi, S. U. Rawal, and M. M. Patel, External stimuli-responsive drug delivery systems, in Drug Delivery Devices and Therapeutic Systems, Elsevier, 2021, pp. 267–288. doi: https://doi.org/10.1016/B978-0-12-819838-4.00023-7
166. N. Dubey, R. Bentini, I. Islam, T. Cao, A.H. Castro Neto, V. Rosa, Graphene: A versatile carbon-based material for bone tissue engineering. Stem Cells Int. **2015**, 1–12 (2015). https://doi.org/10.1155/2015/80111114213
167. A. Rajan Unnithan, A.R.K. Sasikala, C.H. Park, C.S. Kim, A unique scaffold for bone tissue engineering: An osteogenic combination of graphene oxide–hyaluronic acid–chitosan with simvastatin. J. Ind. Eng. Chem. **46**, 182–191 (2017). https://doi.org/10.1016/j.jiec.2016.10.029
168. P. Zare, M. Aleemardani, A. Seifalian, Z. Bagher, A.M. Seifalian, Graphene oxide: Opportunities and challenges in biomedicine. Nano **11**(5), 1083 (2021). https://doi.org/10.3390/nano11051083

Chapter 2
Graphene-Enhanced Supercapacitors and Functionalisation

2.1 Introduction

Presently, electrochemical energy storage systems (EESS) have a critical role to play in every individual's life and energy generation industry. These devices have found ample applications ranging from portable and microelectronic electronic instruments to power grids as well as electric vehicles [1]. In this regard, from the past decades, researchers have shifted their focus in searching for better electrochemical devices that can deliver prolonged life cycles, are cost-effective, and also offer better power as well as superior energy densities [2]. Among the several types of energy-storing devices, supercapacitors have gained great attention owing to their ability to supply high electrochemical energy performance [2].

Further, to design supercapacitor electrode materials with superior performance necessitates the incorporation of nanomaterials, which eventually improves their pseudocapacitance. To date, numerous nanomaterials, like conducting polymers, reduced graphene oxide (rGO), activated carbon, and 1D nanotubes of carbon (CNTs), have been researched to be included in supercapacitor electrodes due to their ability to provide expanded surface area, unmatched physicomechanical traits. Thereby enhancing the electrochemical energy measures synergistically. Among the various allotropic forms of carbon, graphene has been regarded as the most significant carbonaceous substance for storing energy due to its incomparable physical, mechanical, and chemical characteristics. It has exhibited outstanding power of 130 GPa, superior charge transfer of 200,000 cm^2V^{-1} s^{-1}, and thermal conductivity of 5300 Wm^{-1} K^{-1}. The large surface area, high chemical stability, and tailorable surface functionalities make graphene a highly sought material for energy storage applications [3].

V. Khanna, C. Prakash, *Graphene for Electrochemical Energy Storage*, Lecture Notes in Nanoscale Science and Technology 24,
https://doi.org/10.1007/978-3-031-87643-1_2

2.2 Overview of Supercapacitors

Conventional electrochemical energy storage systems are not only toxic from an environmental perspective, but they also lag in terms of technological advancements. Iron-based energy-storing devices exhibit longer life, are effective, and are eco-friendly. However, the energy density of these devices is lower in aqueous solutions [4]. This has prompted the researchers to shift their focus on developing devices which can deliver enhanced electrochemical performance with larger capacity and longer lifespan. Considering these aspects, the use of supercapacitors has become an alternative solution for energy-conserving applications [5]. This is because supercapacitors can engender upgraded power density (Fig. 2.1a), swift charging-discharging period, better life cycle, and most importantly, they are less environmentally damaging [6].

Considering the distinguished methods for storage of energy, supercapacitors are categorised into three groups: hybrid supercapacitors, electrochemical double-layer capacitors (EDLCs), and pseudocapacitors, as depicted in Fig. 2.2b. While the working mechanism in EDLCs is established on the non-faradaic method, the pseudocapacitors work on faradaic reactions. The hybrid supercapacitors make use of both the faradaic reactions and the non-faradaic reactions to store energy [7–9]. The major component of supercapacitors, which have a critical function to play in the storage mechanism, is the electrode. Different types of electrodes, for example, supercapacitors make use of metal and non-metal oxides as well as carbon-based materials depending on the working process mechanism. Besides this, electrolytes have a substantial influence on the supercapacitor's potential window. The impact of electrolytes, in particular aqueous, organic, and ionic, on supercapacitor functioning has been studied.

2.2.1 Electric Double-Layer Capacitor (EDLC) Supercapacitor

In 1853, the first model of EDLC was designed by von Helmholtz, which has a close resemblance to conventional capacitors. Later, it was amended by Gouy and Chapman [12]. Figure 2.2 illustrates the fundamental concept of an EDLC. It consists of cathode and anode electrodes that are submerged in an electrolyte. The two phenomena, namely desorption and adsorption, are employed to store the electric energy in the form of a two-layered charge structure forming across the interface of the two electrodes. Also, prevents ion or electron exchanges between the electrolyte solution and electrodes [13]. During charging and discharging, electrons transfer via the external circuit. Both the positively charged cations and the negatively ionic charged anions would migrate in the direction of the negative electrode, leading to the electrostatic separation of charges. The ions then undergo a desorption process and return to their original electrode during the discharging cycle. The ion's movement thus creates an electric double layer that stores charge [14].

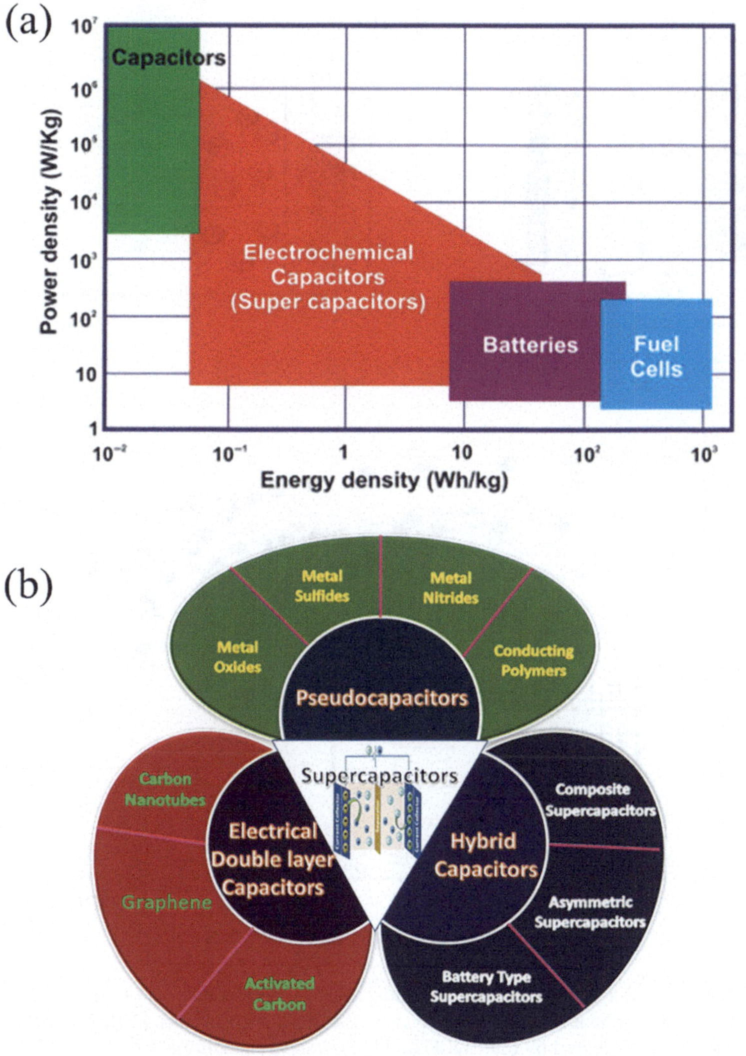

Fig. 2.1 (**a**) Graphical representation provides a comparison of both the power and energy densities for distinctive energy collecting devices [10] and (**b**) different electrode materials utilised in the double-layer capacitors, hybrid capacitors, and pseudocapacitors [11]

Both the volumetric and gravimetric capacitance, in general, are higher in EDLCs, which is led by the vast surface area established with the electrode materials. Also, electrode materials enable the EDLCs to exhibit prolonged cycle life [13]. The electrodes usually swell as the battery is charged or discharged. However, due to no faradaic reaction in the EDLCs, the swelling is hindered. Moreover, to ensure the ionic transfer of ions among the two electrodes, both organic and aqueous electrolytes can be utilised. Still, for storing more charges, organic electrolytes are mostly preferred over aqueous ones [17].

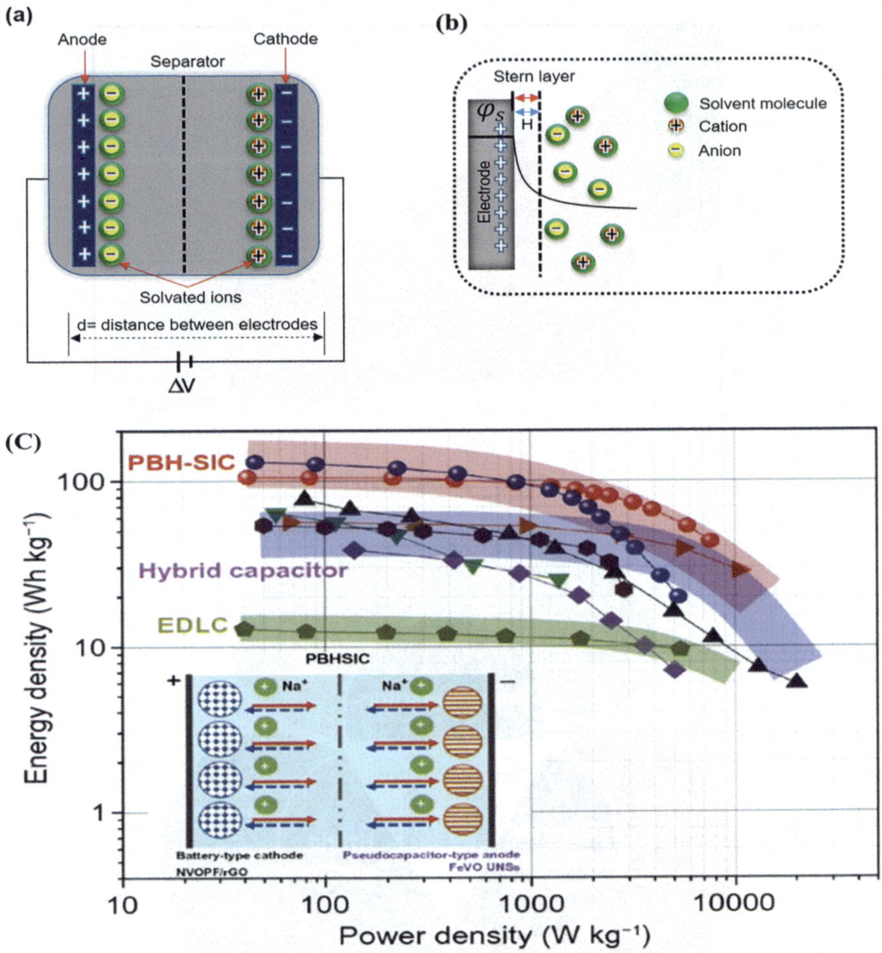

Fig. 2.2 Schematic diagram of (**a**) electrochemical EDLC supercapacitor composed of two electrodes and (**b**) development of Helmholtz layer on the electrode's surface [15]. A comparison distinguished densities delivered by the EDLC and hybrid supercapacitors [16]

Moreover, to enhance the performance of EDLCs, the electrodes are prepared using highly porous carbon materials, to name a few activated carbon [18], one-dimensional carbon nanotubes [19, 20], carbon aerogels [21], 2D graphene [22], and carbon foam [20]. The use of carbonaceous materials allows EDLCs to exhibit outstanding charge-discharge cycle stability along with higher retention capacity [20]. They also show outstanding power density and enhanced specific capacitance, but the energy density is relatively very poor due to no-faradaic processes. Among all the three supercapacitors, EDLC shows the lowest density, as shown in Fig. 2.2c.

2.2.2 Pseudocapacitors

The word pseudocapacitors originated from the early Greek term where pseudés means false. In context to supercapacitors, it means not exactly but appearing to be. Conway utilised the term pseudocapacitance for exhibiting materials, which tend to show electrochemical behaviour much better compared with EDLCs [23]. In pseudocapacitors, the mechanism is faradaic. Storage of energy is quite rapid [24]. In pseudocapacitive materials, a mechanism for the storage of charge could be through intercalation reactions or redox reactions near the electrode surface (Fig. 2.3).

In redox reaction pseudocapacitors, the storage of charge is primarily ascribed to the ions moving across the surfaces of the electrode [24]. They make use of different methods, namely double layer and surface faradic mechanisms for charge storage. Such materials which offer remarkable innate electrochemical characteristics are referred to as pseudocapacitive materials [26]. In the former type of pseudocapacitors, few materials which are arranged in layers may experience faradaic charge movement owing to electrolyte ion intercalation with layers without any crystallographic change of phase [27]. They are well-known for generating not only reversible but also rapid rates for charge storage, which can also exceed the near-surface redox pseudocapacitors [26].

Pseudocapacitive materials that have a vast surface area, like graphene, can attain fast charging and discharging of pseudocapacitors. Since the slow diffusion does not impede their charge-storing means, they are mechanically distinct from batteries [28]. Nonetheless, pseudocapacitors are highly susceptible to chemical reactions owing to the faradaic charges. This can decline their electrochemical performance [24, 28]. Aerogels are another alternative material which exhibits the great capability to be used as energy storage electrodes for their super lightweight, porous architecture, and superior electrical conductivity [29]. Other than graphene and aerogels,

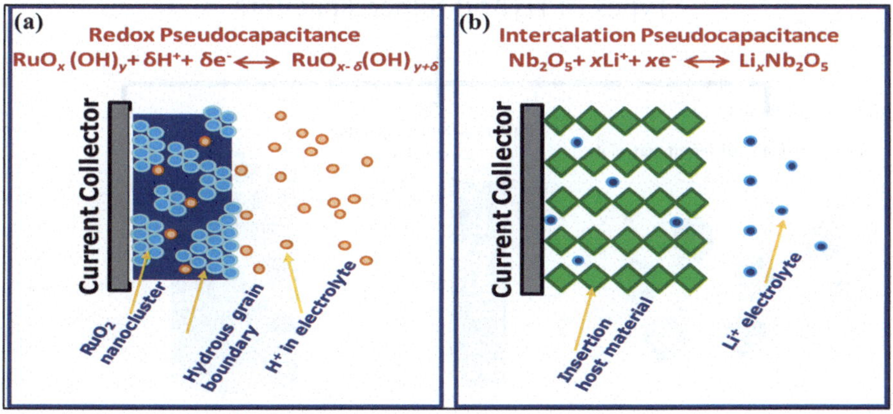

Fig. 2.3 Illustrates (**a**) the chemical process for generating pseudocapacitance in (**a**) surface redox pseudocapacitor and (**b**) intercalation pseudocapacitor [25]

transition metals with oxides and hydroxides are also utilised in pseudocapacitors because of their persuasive electrochemical characteristics [30].

2.2.3 *Hybrid Supercapacitor*

Hybrid supercapacitors are specifically categorised as asymmetric supercapacitors, battery-based hybrid supercapacitors and self-charging supercapacitors. In the asymmetric and hybrid battery supercapacitor, both the electrostatic adsorption and redox reaction mechanism are applied to simultaneously increase the specific power and energy, respectively. The asymmetric supercapacitor utilises two distinct electrodes, one from the battery type and the other from the supercapacitor, as shown in Fig. 2.4.

Asymmetric supercapacitors can attain higher operating voltage as well as high specific capacitance. This is due to the voltage widow deviation engendered by the different electrode materials, which can enhance the energy density substantially. Electrode material hybridisation also has great promise for energy storage. The combination of more than two materials together leads to the formation of a hybridised supercapacitor. Hybrid supercapacitors that are inspired by batteries combine the usage of batteries with supercapacitors. The self-charging type supercapacitors can combine and store different energy sources simultaneously [31]. Mainly, there are two factors, namely electrode and electrolyte, that contribute to the higher performance of asymmetric supercapacitors. Different material combinations can be incorporated together to improve the capacity of the supercapacitors under a suitable electrolyte solution.

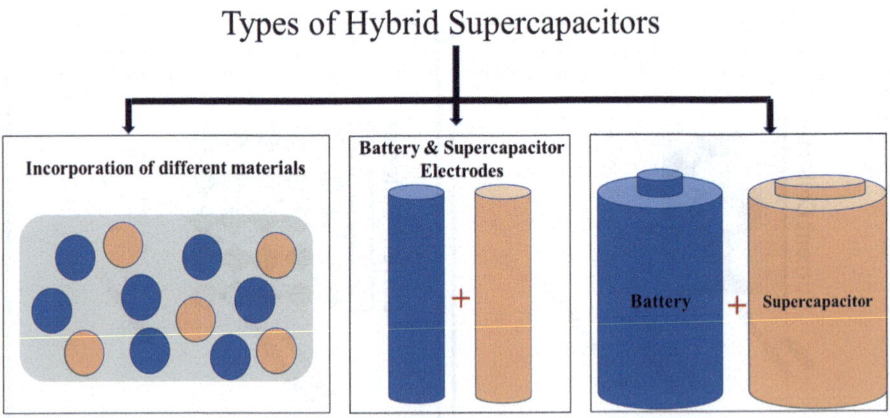

Fig. 2.4 Schematic representation of different types of hybrid supercapacitors can be developed by combining different materials, different electrodes, and battery and supercapacitor combinations

Lately, hybrid battery supercapacitors have increasingly sought attention on account of their improved electrochemical behaviour, cost-effectiveness, and safe operation [32, 33]. The inclusion of batteries and supercapacitors together to form a hybrid supercapacitor has the advantage that it can simultaneously increase both the specific energy and specific power significantly. In other words, the formation of a hybrid battery supercapacitor is far better than using individual technology serving a limited purpose.

2.3 Functionalisation—Tailoring Graphene for Performance

Electrode materials pose numerous challenges which critically influence the electrochemical kinetics of the supercapacitors. The most common concern with the electrodes is to improve the ionic conductivity. It is highly desirable for escalating the performance rate as well as the energy density. Some oxide electrodes (anhydrous RuO_2), for example, exhibit poor ionic conductivity as compared to their counterparts ($RuO_2 \times H_2O$) and limit the application of supercapacitors [34–36]. Besides this, poor conductivity revealed by the electrodes causes the supercapacitor's functionality to deteriorate. To illustrate, nickel oxide-based electrodes in a supercapacitor exhibited poor cyclic stability and lower specific capacitance due to its minimal electrical conductivity [37]. Another major issue related to electrodes is the poor surface area, which hinders the accessibility of ions on the surface. Apart from this, another limiting factor is the lower mechanical strength and electrochemical stability. Volume changes in the electrode during charging and discharging can damage the electrode due to poor mechanical strength of electrode. Moreover, functionalising the graphene structure can further enhance its capability to boost the electrochemical productivity of the electrochemical devices.

To address these concerning challenges, graphene, a carbonaceous material, has been gaining great traction from the research and commercial community. The astonishing innate features of graphene are the reason it has become a highly suitable material for the development of supercapacitor electrodes. Specifically, graphene may be functionalised by adding compounds to the edge or basal planes or by undergoing π–π stacking interaction, as shown in Fig. 2.5.

From the previous studies, it can be established that synthesis of graphene free from oxygen content is quite difficult to achieve as the suspended carbon molecules are capable of absorbing the moisture easily. As the oxygen content is relatively low in content, it does not add to its hydrophilic and pseudocapacitive nature. To enhance these properties of graphene, it is thus desirable to increase the oxygen content on its surface. Additionally, after doping, the graphene emerges with a substantial quantity of oxygen content. And this amount increases with the increasing concentration of doping within the graphene surface.

To increase the active sites to attach or graft different functional groups, ribbon-like morphology with additional edges is highly useful. As shown in Fig. 2.6a, thermal reduction and chemical synthesis methods can be performed to attain

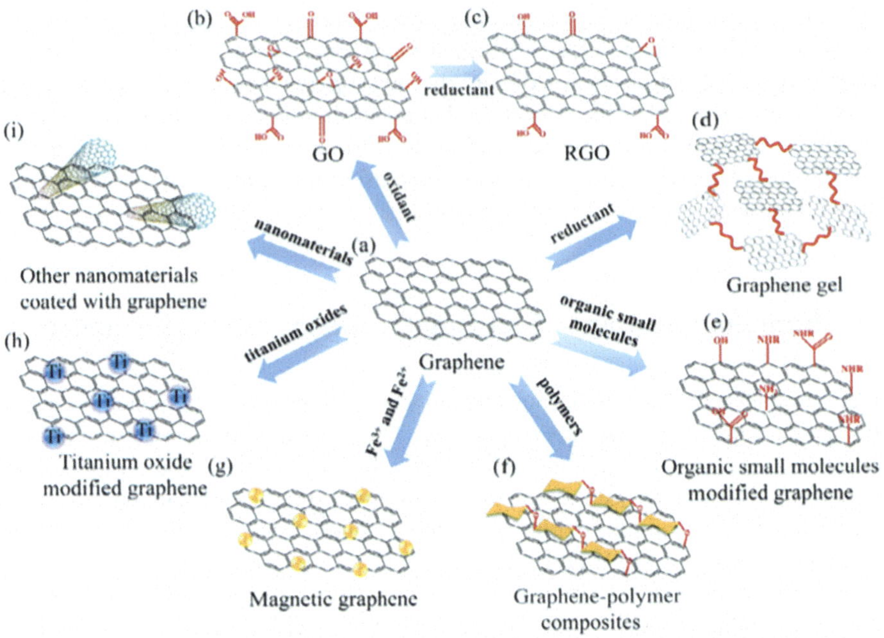

Fig. 2.5 Demonstrates different methods of functionalising graphene structure with various materials and techniques on edge and basal planes [38]

ribbon-structured oxygen and sulphur-functionalised graphene. It was observed that for a 1.0 $mgcm^{-2,}$ the oxygen-functionalised and sulphur-functionalised graphene depicted specific capacitance values of as high as 428 and 1660F/g for the current density of 1 A/g, respectively. Further, an increase in mass loading to 10 mcm^{-2} still provided the specific capacitance with a high value of 300 and 935 F/g again for the current density of 1 A/g. Both the electrodes also delivered excellent energy density along with the extended power density of 106.6 Wh/kg and 101.4 kW/kg. Moreover, the supercapacitor revealed 100% retention for over 20,000 cycles with remarkable stability Fig. 2.6c. The reason for this dramatic achievement is the conductive framework produced by the graphene surface for enhanced electron transfer and functional groups caused superior specific capacitance through faradaic redox interactions [39]. Likewise, in another study, different variants of graphene fabricated under varying temperature regimes with several oxygen groups on their surface showed a higher retention capacitance of nearly 98% with an optimised lifespan of 10,000 cycles [40].

Interestingly, using a single-pot technique, the tris aminomethane (T) functionalised into the graphene oxide (GO) worked as a supercapacitor electrode. The structural reactions and appearance of the T functionalised GO are depicted in Fig. 2.6d, g. The developed electrode had shown exceptional capacitance of 549.8 F/g for a current density of 2.5 A/g. While at 7 A/g, the capacitance per unit mass attained was 358 F/g. It also retained 80% of initial capacitance for over

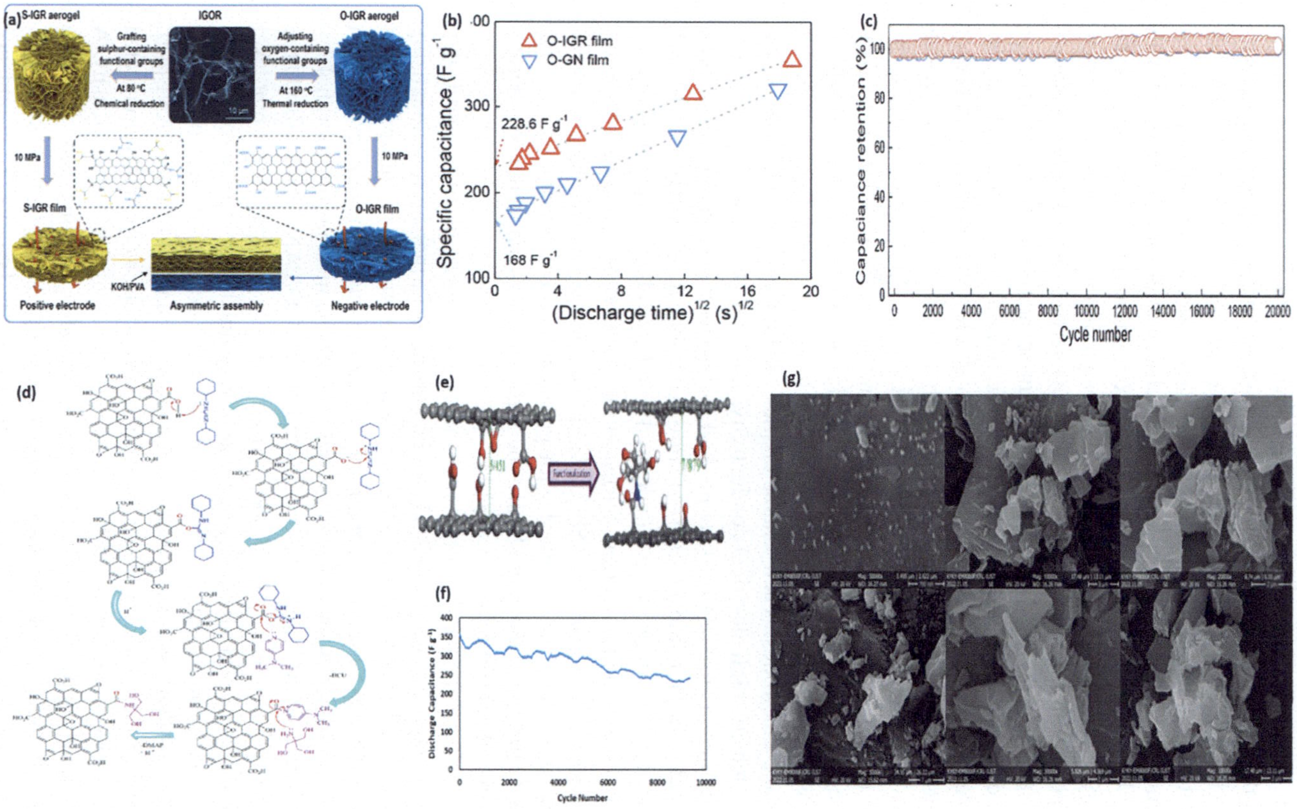

Fig. 2.6 Schematic representation (**a**) steps to synthesise sulphur and oxygen-functionalised graphene aerogel using two different methods, (**b**) comparison between OIGR and OGN films in terms of specific capacitance concerning discharging period, (**c**) retention of capacitance of OIGR over 20,000 cycles [39], (**d**) structural model of functionalised graphene, (**e**) difference in interlayer distancing between graphene layer before and after functionalisation, (**f**) attained capacitance after graphene functionalisation, and (**g**) SEM images of pristine graphene, S-IGR, and OIGR electrodes [41]

5500 cycles and 68% retention for over 9300 cycles. The theoretical calculation model depicted the spacing between the layers of GO was increased after functionalisation, as shown in Fig. 2.6e [41]. This enhancement in capacitance can be ascribed to the hindrance to restacking, along with the increasing the specific surface area, which would have increased the charge storage capacity of the double-layer capacitor as additional area is now accessible to electrolytic material [42].

Furthermore, in an investigation, GO was developed after functionalisation with Arginine to be applied as an electrode for supercapacitor application using green methodology at room temperature. Due to the design of a three-dimensional interconnected porous morphology, alteration of the GO surface chemistry, and expanding the interlayer spacing, the multi-functional Arginine makes the GO supercapacitor unique and dramatically increases its performance. With these structural characteristics, the Arg/GO–SSC–device was designed to attain a superior capacitance of 295 Fg/1 for a specific current at 1 A/g, along with the 94% cycle stability for above 4000 cycles in an electrolyte solution of KOH (6 M). In addition to this, the remarkable densities of energy of 50 Wh/Kg and power 2204 Wkg of the Arg/GO–SSC–mechanism demonstrated its remarkable energy storage capabilities [43].

In contrast to stacked graphene sheets, aerogels have a substantial capability as energy storage electrodes. Their characteristics, including their extremely lightweight, low density, high electrical conductivity, and three-dimensional porous network, are the main cause of this. One method for producing aerogel is to 3D print graphene oxide (GO) ink into a lattice structure. After that, the material goes through further procedures to improve its conductivity and usefulness, including drying, annealing, oxidation, and reduction. Surface-functionalisation followed by the creation of aerogels of 3D graphene (SF-3D GA) is a noteworthy accomplishment of this method. The electrode designed with graphene aerogel provided high areal capacitance (2195 mF × cm^{-2}) with also the intrinsic charge capacitance reaching the maximum value (309.1 μF × cm^{-2}) [44].

Also, the electrochemical energy performance of supercapacitors fabricated utilising the composites of rGO/PANI can be enhanced further by functionalising rGO with amines. The outcomes of a study revealed that altering the morphological features of F-rGO/PANI increases their specific capacitance. The enhancement in energy density and cycling stability is associated with the mesoporous morphology of F-rGO. The specific capacitance delivered by F-rGO/PANI charge/discharge electrode was 597 F/g, and by PF-rGO/PANI was 498 F/g at a specific current level arranged to 0.5 A/g. Both these were comparatively much higher than for pristine PANI. Also, the retained capacitances for both of the developed electrodes were 75% and 89% of their initial capacitance, far greater than that of pure PANI [45]. Furthermore, compared to conventional graphene materials, a covalently connected graphene framework containing micropores may offer a greater energy density. A unique graphene network was formed by covalently bonding the graphene nanosheets with anthraquinone pillars. Compared to a pristine graphene electrode, the functionalised graphene electrode structure has many benefits. To commence with, functionalising inhibits the restacking of the graphene nanosheets since the covalently bonded architecture is strong. Additionally, the length of the pillars may be used to adjust the spacing between graphene sheets, enhancing specific capacitance through the nanopore effect. Also, the area of the specific surface of graphene sheets is further increased since electrolytes may reach both sides of the sheets. Last but not least, the pores within the structure are coupled to empower swift mass and charge transfer, providing the possibility of high-power density and retention rate. The F-rGO electrode in the capacitor maintained 200 F/g of specific charge capacitance across a specific current of 100 A/g. Evident the exceptional deliverance of power density, also with considerable rate capability. Also, the CV profile at a 50 mV/s scan rate under 1 M H_2SO_4 demonstrated the high stability of the F-rGO electrode. Over 5000 complete cycles, 96% of capacity retention was attained, exhibiting high specific capacitance [46].

From the above-mentioned experimental studies, the graphene as a whole electrode material has limited utilisation as it may lag in different zones of electrochemical energy measure. However, the functionalised graphene-assisted electrodes in supercapacitors observe great transformation in energy storage applications. This is mainly due to the enhancement in the active sites, increased interlayer distancing and vast areas of specific surface that develop high power and energy densities in the supercapacitors.

2.4 Graphene's Role in Electrochemical Capacitors

In present times, establishing a safer and more sustainable energy-efficient future is one of the most pressing social as well as scientific issues. In this regard, electrochemical capacitors can play a significant role towards a sustainable future as an electrochemical energy storage system. However, their energy-storing techniques are quite rapid at the surface or near-surface, they offer excellent power density in trade-off with energy density. Henceforth, it is highly sought to fabricate electrochemical capacitors with high electrochemical energy measures, specific power and specific energy along with rate capability [47].

Lately, constant research in graphene, offering high physicochemical properties, based electrodes has gained substantial interest worldwide. As a result, composite materials based on graphene have been explored for use in several EES systems. These composites are enabled to make use of the benefits of both the graphene nanosheets and electrochemical materials to provide improved electrochemical energy performance. However, difficulties in clearly understanding the EES mechanisms of various graphene composites also appeared as a result of the intricacies of their microstructures and electrode processes. These challenges demand systematic electrochemical analysis by sufficient theoretical evaluations and sophisticated operating practices [48].

In the duration of the cycling period, the graphene within the specifically developed graphene-based composites is crucial in attracting and transferring electrons to and from the electrochemically active materials. Therefore, reducing the resistance of interface among the dissimilar materials also depends on the attainment of optimal interfacial contacting among the graphene nanosheets and the electrochemically enhanced materials [49].

The composites fabricated with graphene can primarily be classified into different categories (Fig. 2.7a) based on the electrochemical material aspects concerning dimensions and the flexibility of graphene sheets. Porous structure, electrolyte with better wettability and modifying the morphology may be applied for elevating the interface interaction amongst the graphene-based composite and the electrolytic solution. For incorporating porous structure into the composite, different methods can be effectively applied, such as etching holes in the graphene structure, forming aerogels and creating pores during phase two [50, 51]. The most widely employed method among them is forming graphene aerogel structures, which is known to have benefits for creating electrodes with 3D porous materials, including extensive porosity, enlarged surface area, and strong electrical conductivity [52]. Ultimately, it offers an effective porous architecture for forming composite electrodes based on graphene with a three-dimensional ion transfer pathway and ultra-high electrical conductivity.

Enhancing the interactions among the electrolyte and electrode composites of graphene is an additional essential instrument for quickening the dynamic mass transfer process for energy storage processes. Numerous efforts have been applied by researchers to increase the interactions among the conductive network of graphene and the different electrolytes, such as adding foreign atoms, altering the graphene morphology with the incorporation of functional groups and including defects [55–57]. Also, the positioning of the nanosheets of graphene within the composite in a different direction can aid in enhancing electrolyte infiltration. In other words, graphene components within the composite can be positioned majorly in three systems: randomly, aligned parallelly, and oriented vertically. The vertical orientation of the graphene nanosheets is a unique strategy for ion diffusion since several active ions cannot flow through the defect-free graphene planar structure [1].

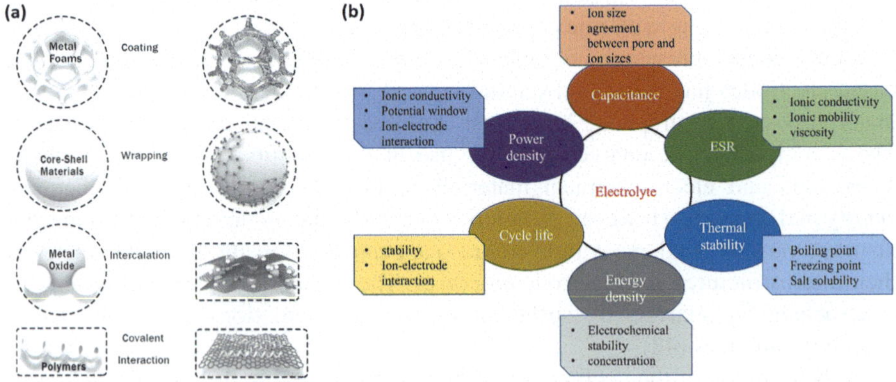

Fig. 2.7 Illustrates (**a**) different forms of graphene composites for supercapacitor electrodes [53] and (**b**) numerous parameters that have a strong influence on the performance of an electrolyte in a supercapacitor for energy-storing applications [54]

The electrochemical behaviour of capacitors has been observed to be far superior with aqueous electrolytes than with a non-aqueous solution. They offer superior power densities in comparison to organic electrolytes. These electrolytes can be alkaline, acidic, or even neutral [54]. The potential window of neutral electrolytes is nearly two times of acidic and basic electrolytes. Different aspects that affect the performance of electrolytes are shown in Fig. 2.7b. Conversely, organic electrolytes are widely used in industries for supercapacitor fabrications although they exhibit lower electrical conductivity. The major drawbacks of organic electrolytes for electrochemical capacitors are their toxicity, low conductivity, and low capacitance [54]. The ionic liquids are emerging as an indispensable part of electrochemical capacitors. They have melting points ranging below 100 °C and are made of ions. Liquid ionic materials exhibit better thermal stability and higher operative voltage and are non-flammable [53].

The reversible reactions occur at the electrode's surface proximity of active materials, rather than ion diffusion occurring at the inside of the active electrode substance; electrochemical capacitors have shown the potential to reach substantially greater power densities than rechargeable batteries. As a result, electrochemical energy capacitors appear to be a good addition to and substitute for rechargeable batteries. Nevertheless, their low cell voltage and restricted energy density, which come from reactions on the electrode surfaces, prevent their extensive utilisation in energy storage technologies. Therefore, to gain high specific capacitance and energy density, electrochemical capacitors are developed with advanced technologies such as using one-, two-, or three-dimensional nanocarbon materials. The two-dimensional graphene offers a larger surface area to accelerate the charge movement by reducing the size of the electrodes [58].

2.5 Graphene in Pseudo and Hybrid Supercapacitors

Graphene supercapacitors are considered well-meaning devices to generate far greater power densities and energy densities enhancing the capacitors' electrochemical performance and charge-discharge cycle rate. The primary objective of supercapacitors is to attain larger charge storage capacities. They are made up of electrolytes, separators, current collectors, and high-surface electrode material. When considering supercapacitors' possible industrial applications, one of their most important features is their increased specific capacitance. Nevertheless, there are several barriers in the way of their actual viability. As a result, graphene-carbon nanomaterials, as well as their derivatives, are expected to enhance their electrochemical performance on a larger scale.

Graphene oxide, functionalised graphene, and their composites are examples of graphene derivative-based electrodes. Chemical synthesis is used to derive these materials. These materials depict excellent electrochemical energy performance characteristics. For example, strong capacitance properties are shown by graphene sulphate (GA) in 1 mol/L sulphuric acid. GA electrodes show remarkable stability

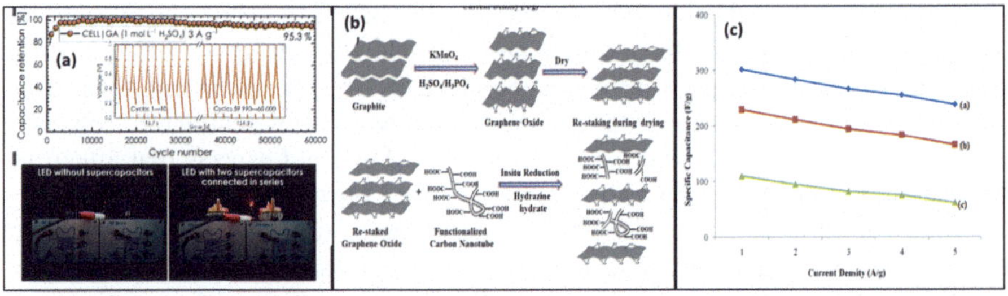

Fig. 2.8 Shows (**a**) the graphene-based electrode's cycling performance over 60,000 cycles. An in-depth look at the first and last 10 cycles between −0.35 and −0.65 V may be seen in the inset [59], (**b**) the number of steps followed in synthesising the rGO/FCNT electrode composite for energy storage application in a supercapacitor, and (**c**) a graphical plot of specific capacitance [62]

in dual electrode batteries, with a 95.3% capacitance rate for more than 60,000 cycles of charge and discharge for a value of specific current of 3 Ag^{-1}, indicating its substantial competence for long-lasting performance in harsh environments (Fig. 2.8a) [59]. In a different study, at a modest value of current density 327 Fg^{-1}, the rGO electrodes with functionalised graphene produced a greater specific capacitance of approximately 327 Fg^{--1}. The electrode also showed considerably enlarged power density with a decent energy density (46.96 Wh Kg^{-1}). Its remarkable stability was demonstrated by the fact that it maintained 87.8% of its initially obtained capacitance for more than 5000 fast cycles. This green reduction process presents a viable way to produce rGO with improved electrochemical characteristics, which makes it a unique carbon-structured material for supercapacitor hybrid electrodes [60].

Using the spinning process, an alginate composite loaded with Ni in a flexible architecture containing GO nanofibres was created. 150–500 nm diameter hollow CNT/RGO composite fibres with 366.98 m^2/g of broadened surface area were produced when Ni functioned as a catalyst. The results of the electrochemical examination showed that the symmetric and EDLC capacitors had greater capacitances (267.8 and 49.3 F/g). Ultimately displaying their increased specific surface area and potential as electrode composites in pliable and versatile energy-storing devices [61]. Similar to this, H_2SO_4/HNO_3 was utilised in combination with CNTs to introduce carboxylic groups. At 302 F/g of capacitance per unit mass and a specific current of 1 A/g, the resultant functionalised CNT added into graphene showed great promise as supercapacitor electrodes [62].

Moreover, a composite polymer with a spongy structure was formed to enhance the effectiveness of supercapacitors using a one-step hydrothermal process with activated carbon as well as graphene as the origin elemental materials. The resulting conductive hydrogel-based electrode exhibited 26.9 Whkg-1 of density (energy) at 242 Wkg-1 and an astounding measurement of capacitance of 505.6 Fg^{-1}. Activated carbon doped with nitrogen along with graphene was used to form the electrode. Its capacitance retention of around 92.1% after over 5000 cycles is noteworthy and confirms its prospective application in supercapacitors [63]. Furthermore, lignin-based biochar was included within the graphene by facile and rapid co-precipitation synthesis approach for producing the composite. For a potential rate of scan at 20 mVs^{-1}, the developed material displayed normal electrochemical impedance having a charge capacitance per unit mass of 103 Fg^{-1} [64].

Graphene has been successfully used to add metal oxides, including oxides of vanadium (V_2O_5), lanthanum La_2O_3, molybdenum (MoO_2), and iron (Fe_2O_3), on its surface by using the microwave-yielding technique. This method has made it possible to improve certain capacitances in a range of supercapacitor applications. For example, at 1 Ag^{-1} current charge density, an exceptional initial capacitance of charge reaching 673.2 Fg^{-1} was recorded, which at 10 Ag^{-1} had shown better retention capacitance of 476.6 Fg^{-1}. Additionally, the synthesised materials showed remarkable cyclic stability, retaining 96.8% of their original value for more than 10,000 complete electrochemical cycles at a specific current of 1 Ag^{-1}. Indicates the robustness and longevity of the metallic oxide's electrode architectures based on graphene in energy storage devices. Metallic oxides are more evenly deposited on the surface of materials using the microwave synthesis process [65].

Using electrochemical exfoliation based on plasma, manganese oxide/graphene electrodes are produced as an additional tactic. For a potential scanning rate of 10 mVs^{-1}, this approach produces a charge capacitance of ~320 Fg^{-1}, indicating that the single-step methodology is a workable, economical, and environmentally benign procedure. In a later investigation, graphene was additionally incorporated with a 3D MnO_2 fibre architecture at the interface. A high specific charge capacitance (683 Fg^{-1}) for a specific current of about 2.2 Ag^{-1}, along with a remarkable cycling performance of 96.9% retention initial capacitance, was reached for more than 5000 complete cycles with GNS-MnO_2 composite. Using an alternative method, the ESS's electrochemical behaviour was enhanced by creating a three-dimensional composite material. This material blends MnO_2 with porous rGO. A noteworthy charge capacitance per unit mass of 192.2 F/g was attained using this material at 0.5 Ag^{-1} [66].

Graphene-PANI nanocomposites have shown great promise as supercapacitor electrode materials in recent years [67, 68]. The composite material made of rGO/PANI flakes showed 1182 Fg^{-1} of impressive capacitance when used in a combination of a three-electrode configuration at 1 A/g and 808 F/g when used in a two-electrode composition with symmetry.

A substantially improved value of energy density reaching up to 28.06 Wh/kg was also attained by the rGO/PANI composite at a higher range of power density reaching 250 W/kg [69].

Recent research has investigated several methods to improve the functional capacity of supercapacitors composed of electrodes that contain rGO and PANI. Porous PANI/rGO composites, for example, that were hydrothermally synthesised, showed a range of specific capacitance that varied from 420 to 239 F/g, spanning current densities of 0.2–20 A/g. For over 6000 cycles at a specific charge/discharge current of 2 A/g, this composite showed remarkable stability, holding 80% of its capacitance [70]. Moreover, 3D graphene was introduced with polyaniline (foam material) and doped with nitrogen to form N-GE/PANI, which exhibited exceptional electrochemical performance, retaining 95.9% capacitance with more

Table 2.1 Comparison of the electrochemical energy measures for supercapacitors with electrode materials synthesised with graphene

Material of the supercapacitor electrode	Material of the current collector	Specific capacitance in Fg^{-1}	Power density in $KWKg^{-1}$	Energy density in $Wh\ Kg^{-1}$	Cyclic stability in percentage retention	Ref.
Boron-doped MnO_2	Fibrous carbon	269 Fg^{-1} at scan rate of 50 mVs^{-1}				[72]
MnO_2 incorporated with GO-CNFs	Nano fibred carbon	612 Fg^{-1} at a scan rate of 2 mVs^{-1}	27.8	30.4^{-1}	Over 5000 complete cycle, 109% retention observed	[73]
$NiCo_2O_4$ without graphene	Used Ni foam	262 Fg^{-1} at current density of 2 Ag^{-1}				[74]
Nanograin MoS_2	Stainless steel	180 Fg^{-1} at scan rate of 5 mVs^{-1}	2.1	23	More than 1000 full cycles, 82% retention observed	[75]
TiO_2	Foil of Ti	13.94 $mFcm^{-2}$ at current value of 0.5 $mAcm^{-2}$	0.51 ($Wcm^{-3)}$	8.5×10^{-4} (Wh cm^{-3})	Over than 10,000 cycles, revealing 90% rate retention	[76]
Graphene-porous α-Fe_2O_3	Form of Ni	343.7 Fg^{-1} at a current density of 3 Ag^{-1}			Complete 50,000 cycles showing a 95.8% of retention capacity	[77]
A-BNQDa incorporate with rGO		256.8 Fg^{-1} at a current density of 1 Ag^{-1}			86.9% retention was observed for more than 5000 complete cycles	[78]
Hybrid of ZnS/rGO/PANI		1045.3 Fg^{-1} at a current density of 1 Ag^{-1}			76.1% retention was noted for over 1000 full cycles	[79]

than 5000 completed cycles. With a significantly greater specific capacitance value, which reached 528 F/g at a measured specific current up to 0.1 A/g, they demonstrated their feasibility to be applied in supercapacitors for high-energy storage [71]. Energy measures of graphene-based composite materials are depicted in Table 2.1.

2.5.1 *Graphene's Role is Pseudocapacitors*

The transport of charged ions along with charged electrons between the electrode and the solution of the electrolyte is the initial step for the operation of pseudocapacitors, which are based on the idea of pseudocapacitance. A consequence of ion adsorption and quick faradaic interaction that takes place throughout cycles is pseudocapacitance. By using the ions to store energy on the surface of the electrode, they produce pseudocapacitance through electrochemical processes. Materials with large surface areas, like graphene pseudocapacitors, have quick charging and discharging times. They are mechanically distinct from batteries because their charge storage process is not impeded by slow diffusion [28]. However, they are weak and can chemically react because of faradaic charges, which might further impair their functionality [24, 28].

Using an electrophoretic deposition technique and a thermally treated single-step process, the Co_3O_4/holey graphene composite demonstrated an outstanding specific charge capacitance value measured to 2714 F/g. With a current charge density varying from 1 to 10 A/g, the composite maintained 78% of its capacity, demonstrating its remarkable rate capability. Additionally, the electrode material had power densities of 0.6 kW/kg in addition to energy densities of 165 Wh/Kg, respectively [80]. Moreover, significant results regarding the electrochemical energy-storing capabilities of doped graphene have also been demonstrated. Recently, the use of N and S simultaneously doped with graphene aerogel for forming NS-3DPGHs material as pseudocapacitor electrodes has been investigated. The co-doped electrode material demonstrated exceptional performance as an electrode. It achieved an enhanced capacitance of 412.9 F/g for a maximum measured specific current of 0.5 A/g and demonstrated a retaining rate of 96.4% of its preliminary charge capacitance over more than 10,000 complete cycles [81].

Among the various energy-preserving devices, the smallest are micro-supercapacitors formed with pseudocapacitive behaviour. They provide numerous prospects for integrating seamlessly into wearable gadgets. Still, the poor voltage generated by aqueous electrolytes is a huge challenge for the electrochemical performance of pseudocapacitors. However, in a study, an aqueous micro-pseudocapacitor was developed, which has an enhanced operating voltage value measured at 1.6 V. Achieving high output was attributed to the synergy effect of the combination of rGO and polypyrrole in the LiCl gel electrolyte of 20 M. The synthesised composite electrode within the LiCl electrolyte delivered an areal capacitance of 44.5 mF/cm^2. as shown in Fig. 2.9a–g. It also exhibited ultra-high mechanical flexibility along with retention capacitance of 97% for 180° of bending angle [82]. Furthermore, MoS_2/rGO hybridised with few-layered MoS_2 nanosheets exhibits a vast specific capacitance (361 F/g) at a rate of charge scan of 5 mV/s. Also, it revealed an excellent rate performance, with long-life stability in an electrolyte solution of 1.0 ML^{-1} H_2SO_4. Extended energy density and great cyclability of 5.2 Wh/kg@62.0 kW/kg as well as 23.0 Wh/kg@0.37 kW k/g are displayed by the symmetric supercapacitor. Due to hybridisation, it results in significantly increased conductivity, more active sites, along more redox kinetics and primarily capacitive behaviour as shown by the in-depth charge transfer kinetics investigation [83]. The electrochemical energy storage performance for pseudocapacitors is depicted in Table 2.2.

The utilisation of a double-layered capacitor process and combining it with the pseudocapacitive materials transforms them into a single hybrid composite to

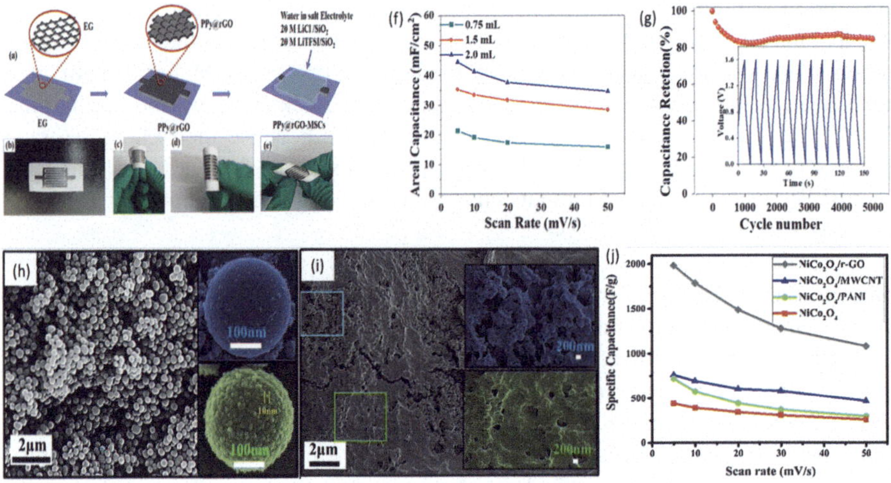

Fig. 2.9 (**a–e**) Steps involved in processing green mini-pseudocapacitor for flexible wearable device based on graphene, (**f**) areal capacitance of green micro-pseudocapacitor (**g**) high retention of initial capacitance over 20,000 cycles [83], (**h**) and (**i**) SEM images of pure $NiCo_2O_4$ and $NiCo_2O_4$/rGO composite, and (**j**) the rate of specific capacitance versus the scan rate [84]

create nanostructured electrodes for supercapacitors, showing feasibility to greatly extend the devices' number of practicable applications and improve their electrochemical performance. In this regard, to form a hybrid mono-structured electrode from distinct materials, the composites were created using simple, affordable, and scalable strategies. To fabricate the hybrid type of electrode, $NiCo_2O_4$ compound served as the foremost reversible fast faradaic-based nanomaterial, while rGO, MWCNTs, or the conductive polymer polyaniline (PANI) exists as the functional nanomaterials in EDLCs. With these designs, the researchers were able to achieve a capacitance-measured value of up to 734 F/g for developed nanostructured composite electrodes of $NiCo_2O_4$/rGO, respectively, at a current density of 1 A/g. Furthermore, MWCNT and rGO-based nickel cobalt oxide electrode composites exhibited exceptional stability, with capacity losses of less than 10% after 10,000 cycles [84].

2.5.2 *Graphene Role in Hybrid Supercapacitor*

A novel but versatile hybrid battery supercapacitor was developed with composite electrodes with substantial surface area electrodes and a solution of electrolyte material containing redox-active vanadium ions. This apparatus consists of two half-cells of rGO added with the hydrogel synthesised in a single step and containing vanadium ions. During charging, the vanadium ions experience oxidation fluctuations, which lead to differences in the half-cell potentials and a significant rise in

Table 2.2 Electrochemical performance measures of pseudocapacitor synthesised using graphene-based materials

Material for the composed electrode	Strategy utilised for synthesis	Quantity and type of electrolyte	Specific or areal capacitance	Retention capacity/ cyclic life	Energy/ power density	Ref.
RGO-based ultrathin MoS_2 electrode material		1 M of L^{-1} H_2SO_4	Attained 361 Fg^{-1} at a scan rate of 5 mVs^{-1}	92% retention was noted for over 10,000 complete cycles	Energy density attained was 23 (Wh kg^{-1}) for a 0.37 (kW kg^{-1}) while for 62.0 (kW kg^{-1}), 5.2 (kW kg^{-1}) was attained	[83]
Mesoporous G based NIO_2	Advance Hummers' technique	2 M of KOH	Reached to 632 Fg^{-1} at 2 Ag^{-1}	99 retention percentage capability for more than 6000 cycles	NiO/G composite with in ASC configuration revealed 30.56 (Wh kg^{-1}) and 2800 (W kg^{-1}), respectively	[85]
Nitrogen-doped rGO incorporated with Mn_3O_4	Single-step hydrothermal process	1.0 M of Na_2SO_4	Reached to 345 Fg^{-1} for a current density of 0.1 Ag^{-1}		22.5 $kWkg^{-1}$ was attained at high current density of 10.0 Ag^{-1}	[86]
Hybrid composite of rGO/Fe_3O_4/ PPy	Polymerisation and chemical process	1 M of KOH	Obtained significant 204 Fg^{-1} specific capacitance at a scan rate 5 mVs^{-1} for a 2.5 Ag^{-1} of specific current		Attained 140.76 (Wh kg^{-1}) gravimetric density along with gravimetric power density of5680 (W kg^{-1}), respectively	[87]
rGO-based $LaNiO_3$ composite	Synthesis via hydrothermal approaches	6 M of KOH	Reached to 920 Fg^{-1} at 1 Ag^{-1}	Attained 99% of cycle stability at current density of 10 Ag^{-1}		[88]

energy density. The developed device doubles the capacity of graphene-based hydrogel supercapacitors without the need for vanadium, achieving a capacity of over 225 mAh/g. The supercapacitor retains around 50% of its capacitance when running on battery power, which may be restored electrically. After 1000 cycles, it can retain 95% of its capacitance [89]. Furthermore, microwave-heated FeS_2

graphene doped with phosphorus revealed encouraging outcomes when utilised as electrodes (Fig. 2.10). At a maximum current density value of 3 A/g, it reached a capacitance per unit mass of 246 mAg^{-1}. Furthermore, at a power density of over 1045 W/kg, the designed composite showed exceptional energy-conserving capacity with over 111 Wh/kg of energy density. The higher energy storage ability is a result of the porous nature of the existing graphene that facilitated fast ion transport, doping, and the conductive network established by the high conductivity graphene [90] (Fig. 2.10h).

Similarly, nickel foam was covered in a variety of CoP/rGO nanosheets to produce the conductive electrode material to be operated in a hybrid supercapacitor with a battery-type structure. The produced electrode demonstrated superior electrochemical performance. The capacity of the CoP/rGO as an electrode was remarkably high at 1438 C/g, much greater than that of the other materials examined. Furthermore, the electrode composite for charge storage upheld 83% of its primary capacitance at an elevated current (10 A/g). This demonstrates the viability of CoP/rGO electrodes for hybrid types of supercapacitors. It's interesting to note that at 1010.5 W/kg power density it displayed 43.2 Wh/Kg of energy density. Additionally, 89% capacitance retention had been observed for more than 10,000 cycles. The absorption energy for the OH− onto the CoP is considerably reduced by rGO, according to theoretical observations, which enhances adsorption capacity and increases electrochemical performance [91].

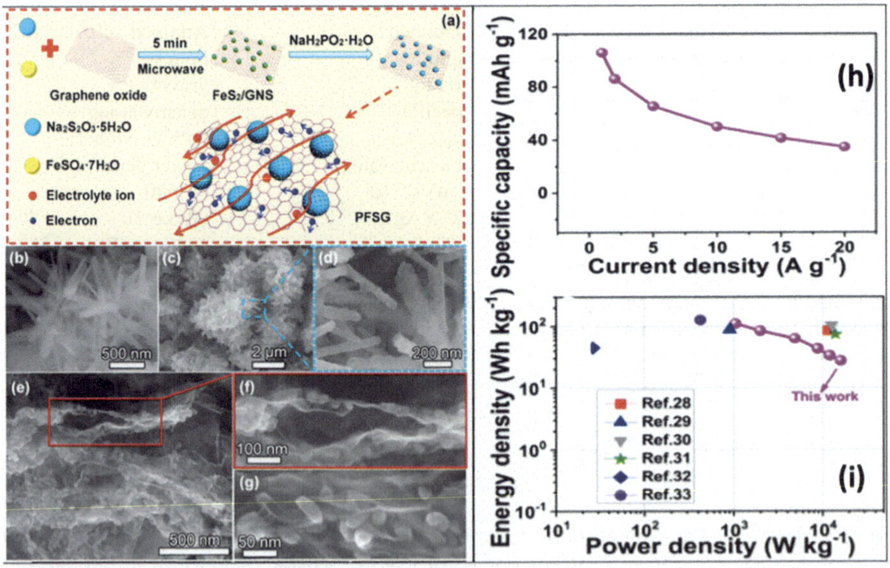

Fig. 2.10 (**a**) shows the structural appearance of pFSG prepared using microwave sintering, (**b**–**g**) SEM images of the FeS_2, P-FeS_2, and PFSG at different resolutions for electrode materials, (**h**) and (**i**) represents the attainment of specific capacitance and energy density by PFSG concerning current and power density [90]

Furthermore, MnO_2/rGO nanoscrolls were coupled with the activated carbon (AC), which served as the cathode to fabricate a hybrid type of energy-storing supercapacitors. GMS/AC-based electrode was formed for hybrid type supercapacitor functioned to close the energy-density gap between supercapacitors and traditional Li-ion batteries (LIBs) with its improved energy density, higher rate performance, and potential Li storage capabilities. The GMS/AC hybrid supercapacitor that was produced has a very high 2040 mAh/g lithium discharge capacity. At its peak, the GMS/AC cell produced a maximum gravimetric power density value reaching 308.1 W/kg and a superior energy density value reaching 105.3 Wh/kg. Additionally, it produced a 42.77 Wh/kg energy density at a power density of up to 30,800 W/kg. Outcomes of the energy density of GMS/AC cells were significantly greater than those of previous graphene-based hybrid devices [92].

One of the main obstacles in the development of the next era of wearables is the creation of green and flexible micro-supercapacitors. Using femtosecond laser pulses and ambient air, it is stated to be possible to directly fabricate extremely conductive, inherently flexible, as well as biologically friendly tiny electrodes from falling petals and leaves without the need for additional materials. Different kinds of leaves are patterned with hierarchically porous graphene using an easy, maskless, and capable scalability laser writing method with a single-step strategy. Biominerals produced from leaves break down and convert into inorganic crystals that act as 3D mesoporous few-layered graphene nucleation sites. In comparison to their synthetic polymer-based equivalents, the FsLIG micro-electrodes from the leaves demonstrate greater areal charge/discharge capacitance (34.68 mF cm^{-2}) at increased specific density (5 mV/s) and reduced resistance among the sheets (23.3 Ω/sq). Also, about 97% capacitance retention was noted for over 50,000 cycles. Small electronic devices, maybe a table watch or a light-releasing diode, can function using such type of hybrid supercapacitor [93].

Additionally, an AC@CC, a combination of AC printed on carbon cloth cathode and an rGO@VO_2 composite anode, was used to prepare electrode components for a hybrid supercapacitor. The rGO@VO_2 composite material facilitates effective redox reactions in supercapacitor applications by having varied V ions (V3+ and V4+) at 0.01–3 V. After more than 120 complete cycles, the 2D to 2D heterointerface structure with V3 ions reaches 1214 mAh/g at 0.1 A/g. Within a charging period of mere 5–834 s, the hybrid battery supercapacitor delivers around 10,000 W/kg power density as well as 126.7 Wh/kg energy density. The rGO@VO2/AC@CC exhibits strong conductivity in battery hybrid supercapacitors, and the hierarchical structure enhances the device's performance [94].

2.6 Conclusion

In this chapter, the latest developments in graphene-based composites, with particular attention to their energy storage mechanisms, interfacial engineering, and rational structural design for various EES applications, have been presented. The mechanism of various types of supercapacitors have also been discussed in detail. The role of graphene in improving the electrochemical behaviour of different energy-storing systems such as EDLC, pseudocapacitors, and hybrid supercapacitors has been described. Functionalising graphene surfaces with different groups and materials for enhancement of electrochemical charge measures of supercapacitors is also recorded. The effect of different electrolytes on the ion diffusion and interlayer spacing between graphene sheets has been demonstrated. Also, the critical part graphene plays in enhancing the surface area of the electrodes for generating more active sites for ion transfer is in this chapter. Recently conducted studies to address the difficulties related to graphene-based pseudocapacitors and hybrid supercapacitors have been detailed. Although significant research has been conducted to augment the electrochemical energy performance of graphene-based supercapacitors, there is still a need to investigate them for further improvement and large-scale production and their real-world applications.

References

1. B. Wang et al., Graphene-based composites for electrochemical energy storage. Energy Storage Mater **24**, 22–51 (2020). https://doi.org/10.1016/j.ensm.2019.08.004
2. M.Z. Iqbal, M.M. Faisal, S.R. Ali, Integration of supercapacitors and batteries towards high-performance hybrid energy storage devices. Int. J. Energy Res. **45**(2), 1449–1479 (2021). https://doi.org/10.1002/er.5954
3. K. Wang, K.N. Hui, K. San Hui, S. Peng, Y. Xu, Recent progress in metal–organic framework/graphene-derived materials for energy storage and conversion: Design, preparation, and application. Chem. Sci. **12**(16), 5737–5766 (2021). https://doi.org/10.1039/D1SC00095K
4. H. Zhang, C. Sun, Cost-effective iron-based aqueous redox flow batteries for large-scale energy storage application: A review. J. Power Sources **493**, 229445 (2021). https://doi.org/10.1016/j.jpowsour.2020.229445
5. W. Jing, C. Hung Lai, S.H.W. Wong, M.L.D. Wong, Battery-supercapacitor hybrid energy storage system in standalone DC microgrids: A review. IET Renew Power Gener **11**(4), 461–469 (2017). https://doi.org/10.1049/iet-rpg.2016.0500
6. M.E. Şahin, F. Blaabjerg, A hybrid PV-battery/supercapacitor system and a basic active power control proposal in MATLAB/Simulink. Electronics (Basel) **9**(1), 129 (2020). https://doi.org/10.3390/electronics9010129
7. W. Du et al., Biological cell template synthesis of nitrogen-doped porous hollow carbon spheres/MnO2 composites for high-performance asymmetric supercapacitors. Electrochim. Acta **296**, 907–915 (2019). https://doi.org/10.1016/j.electacta.2018.11.074
8. D. Wu et al., Morphology controlled hierarchical NiS/carbon hexahedrons derived from nitrilotriacetic acid-assembly strategy for high-performance hybrid supercapacitors. Chem. Eng. J. **433**, 133673 (2022). https://doi.org/10.1016/j.cej.2021.133673
9. Y. Ma et al., Recent advances in transition metal oxides with different dimensions as electrodes for high-performance supercapacitors. Adv. Compos. Hybrid Mater. **4**(4), 906–924 (2021). https://doi.org/10.1007/s42114-021-00358-2
10. S. Sagadevan et al., Fundamental electrochemical energy storage systems, in *Advances in Supercapacitor and Supercapattery*, (Elsevier, 2021), pp. 27–43. https://doi.org/10.1016/B978-0-12-819897-1.00001-X
11. D.K. Singha, R.I. Mohanty, P. Bhanja, B.K. Jena, Metal–organic framework and graphene composites: Advanced materials for electrochemical supercapacitor applications. Mater Adv **4**(20), 4679–4706 (2023). https://doi.org/10.1039/D3MA00523B

12. P. Kurzweil, "Electrochemical double-layer capacitors," in Electrochemical Energy Storage for Renewable Sources and Grid Balancing, Elsevier, 2015, pp. 345–407. doi: https://doi.org/10.1016/B978-0-444-62616-5.00019-X
13. K.A. Sammed et al., Developing next-generation supercapacitor electrodes by coordination chemistry-based advanced functional carbon nanostructures: Progress, current challenges and prospects. Fuel Process. Technol. **250**, 107896 (2023). https://doi.org/10.1016/j.fuproc.2023.107896
14. V.D. Nithya, A review on holey graphene electrode for supercapacitor. J Energy Storage **44**, 103380 (2021). https://doi.org/10.1016/j.est.2021.103380
15. S.W. Bokhari et al., Advances in graphene-based supercapacitor electrodes. Energy Rep. **6**, 2768–2784 (2020). https://doi.org/10.1016/j.egyr.2020.10.001
16. Q. Wei et al., High-energy and high-power pseudocapacitor–battery hybrid sodium-ion Capacitor with Na+ intercalation pseudocapacitance anode. Nanomicro Lett **13**(1), 55 (2021). https://doi.org/10.1007/s40820-020-00567-2
17. S. Sharma, P. Chand, Supercapacitor and electrochemical techniques: A brief review. Results Chem **5**, 100885 (2023). https://doi.org/10.1016/j.rechem.2023.100885
18. G. Fu, H. Li, Q. Bai, C. Li, Y. Shen, H. Uyama, Dual-doping activated carbon with hierarchical pore structure derived from polymeric porous monolith for high performance EDLC. Electrochim. Acta **375**, 137927 (2021). https://doi.org/10.1016/j.electacta.2021.137927
19. B. Pal et al., Exploration of charge storage behavior of binder-free EDL capacitors in aqueous electrolytes. ACS Omega **8**(2), 2629–2638 (2023). https://doi.org/10.1021/acsomega.2c07143
20. P.K. Adusei et al., Fabrication and study of supercapacitor electrodes based on oxygen plasma functionalized carbon nanotube fibers. J Energy Chem **40**, 120–131 (2020). https://doi.org/10.1016/j.jechem.2019.03.005
21. B. Thomas et al., Electrochemical properties of biobased carbon aerogels decorated with graphene dots synthesized from biochar. ACS Appl Electron Mater **3**(11), 4699–4710 (2021). https://doi.org/10.1021/acsaelm.1c00487
22. S.S. Balaji, M. Karnan, P. Anandhaganesh, S.M. Tauquir, M. Sathish, Performance evaluation of B-doped graphene prepared via two different methods in symmetric supercapacitor using various electrolytes. Appl. Surf. Sci. **491**, 560–569 (2019). https://doi.org/10.1016/j.apsusc.2019.06.151
23. T. Brousse, D. Bélanger, J.W. Long, To be or not to be pseudocapacitive? J. Electrochem. Soc. **162**(5), A5185–A5189 (2015). https://doi.org/10.1149/2.0201505jes
24. S. Mahala, K. Khosravinia, A. Kiani, Unwanted degradation in pseudocapacitors: Challenges and opportunities. J Energy Storage **67**, 107558 (2023). https://doi.org/10.1016/j.est.2023.107558
25. M. Sarno, Nanotechnology in energy storage: the supercapacitors, 431–458 (2020). https://doi.org/10.1016/B978-0-444-64337-7.00022-7
26. N.R. Chodankar et al., True meaning of pseudocapacitors and their performance metrics: Asymmetric versus hybrid supercapacitors. Small **16**(37), e2002806 (2020). https://doi.org/10.1002/smll.202002806
27. Z. Liu et al., Constructing enhanced pseudocapacitive Li+ intercalation via multiple ionically bonded interfaces toward advanced lithium storage. Energy Storage Mater **24**, 138–146 (2020). https://doi.org/10.1016/j.ensm.2019.08.026
28. K. Mazloomian, H.J. Lancaster, C.A. Howard, P.R. Shearing, T.S. Miller, Supercapacitor degradation: Understanding mechanisms of cycling-induced deterioration and failure of a pseudocapacitor. Batter Supercaps **6**(8) (2023). https://doi.org/10.1002/batt.202300214
29. L. dos Santos-Gómez, J.R. García, M.A. Montes-Morán, J.A. Menéndez, S. García-Granda, A. Arenillas, Ultralight-weight graphene aerogels with extremely high electrical conductivity. Small **17**(41), e2103407 (2021). https://doi.org/10.1002/smll.202103407
30. M. Khot, A. Kiani, A review on the advances in electrochemical capacitive charge storage in transition metal oxide electrodes for pseudocapacitors. Int. J. Energy Res. **46**(15), 21757–21796 (2022). https://doi.org/10.1002/er.8763

31. D. Gao, Z. Luo, C. Liu, S. Fan, A survey of hybrid energy devices based on supercapacitors. Green Energy & Environment **8**(4), 972–988 (2023). https://doi.org/10.1016/j.gee.2022.02.002
32. D.P. Dubal, O. Ayyad, V. Ruiz, P. Gómez-Romero, Hybrid energy storage: The merging of battery and supercapacitor chemistries. Chem. Soc. Rev. **44**(7), 1777–1790 (2015). https://doi.org/10.1039/C4CS00266K
33. S.M. Benoy, M. Pandey, D. Bhattacharjya, B.K. Saikia, Recent trends in supercapacitor-battery hybrid energy storage devices based on carbon materials. J Energy Storage **52**, 104938 (2022). https://doi.org/10.1016/j.est.2022.104938
34. A. Kumar, H.K. Rathore, D. Sarkar, A. Shukla, Nanoarchitectured transition metal oxides and their composites for supercapacitors. Electrochem Sci Adv **2**(6) (2022). https://doi.org/10.1002/elsa.202100187
35. S. Sahoo, R. Kumar, E. Joanni, R.K. Singh, J.-J. Shim, Advances in pseudocapacitive and battery-like electrode materials for high performance supercapacitors. J Mater Chem A Mater **10**(25), 13190–13240 (2022). https://doi.org/10.1039/D2TA02357A
36. Y. Zhang, Z. Shi, L. Liu, Y. Gao, J. Liu, High conductive architecture: Bimetal oxide with metallic properties @ bimetal hydroxide for high-performance pseudocapacitor. Electrochim. Acta **231**, 487–494 (2017). https://doi.org/10.1016/j.electacta.2017.02.076
37. U.K. Chime et al., Recent progress in nickel oxide-based electrodes for high-performance supercapacitors. Curr Opin Electrochem **21**, 175–181 (2020). https://doi.org/10.1016/j.coelec.2020.02.004
38. A.T. Lawal, Recent progress in graphene based polymer nanocomposites. Cogent Chem **6**(1), 1833476 (2020). https://doi.org/10.1080/23312009.2020.1833476
39. Q. Miao et al., All-functionalized-graphene ribbon films for flexible asymmetric supercapacitors with ultrahigh energy and power densities. Chem. Eng. J. **486**, 150114 (2024). https://doi.org/10.1016/j.cej.2024.150114
40. H. Cao, X. Peng, M. Zhao, P. Liu, B. Xu, J. Guo, Oxygen functional groups improve the energy storage performances of graphene electrochemical supercapacitors. RSC Adv. **8**(6), 2858–2865 (2018). https://doi.org/10.1039/C7RA12425B
41. S. Mohammadi, S.M. Mousavi-Khoshdel, An experimental and computational study of graphene oxide functionalized with tris(hydroxymethyl)aminomethane as an electrode material for supercapacitors. Sci. Rep. **13**(1), 16756 (2023). https://doi.org/10.1038/s41598-023-44048-z
42. M.I.A. Abdel Maksoud et al., Advanced materials and technologies for supercapacitors used in energy conversion and storage: A review. Environ. Chem. Lett. **19**(1), 375–439 (2021). https://doi.org/10.1007/s10311-020-01075-w
43. M. Haghshenas, M. Mazloum-Ardakani, L. Amiri-Zirtol, F. Sabaghian, Arginine-functionalized graphene oxide for green and high-performance symmetric supercapacitors. Int. J. Hydrog. Energy **46**(59), 30219–30229 (2021). https://doi.org/10.1016/j.ijhydene.2021.06.170
44. B. Yao et al., 3D-Printed Structure Boosts the Kinetics and Intrinsic Capacitance of Pseudocapacitive Graphene Aerogels. Adv Mater **32**(8) (2020). https://doi.org/10.1002/adma.201906652
45. Z. Hou, S. Zou, J. Li, Morphology and structure control of amine-functionalized graphene/polyaniline composite for high-performance supercapacitors. J. Alloys Compd. **827**, 154390 (2020). https://doi.org/10.1016/j.jallcom.2020.154390
46. Y. Qin et al., Anthraquinone-functionalized graphene framework for supercapacitors and lithium batteries. Ceram. Int. **46**(10), 15379–15384 (2020). https://doi.org/10.1016/j.ceramint.2020.03.082
47. J. Meng et al., Advances in structure and property optimizations of battery electrode materials. Joule **1**(3), 522–547 (2017). https://doi.org/10.1016/j.joule.2017.08.001
48. C.P. Grey, J.M. Tarascon, Sustainability and in situ monitoring in battery development. Nat. Mater. **16**(1), 45–56 (2017). https://doi.org/10.1038/nmat4777
49. M. Kim, H.M. Hwang, G.H. Park, H. Lee, Graphene-based composite electrodes for electrochemical energy storage devices: Recent progress and challenges. FlatChem **6**, 48–76 (2017). https://doi.org/10.1016/j.flatc.2017.08.002

50. N. Gayathri Devi, T.-H. Mai, R.K. Gupta, P.V. Pham, Art etching of graphene. Nanoscale Horiz **9**(8), 1230–1249 (2024). https://doi.org/10.1039/D4NH00077C
51. K.A.A.A. Elsehsah, Z.A. Noorden, N.M. Saman, Graphene aerogel electrodes: A review of synthesis methods for high-performance supercapacitors. J Energy Storage **97**, 112788 (2024). https://doi.org/10.1016/j.est.2024.112788
52. S. Korkmaz, İ.A. Kariper, Graphene and graphene oxide based aerogels: Synthesis, characteristics and supercapacitor applications. J Energy Storage **27**, 101038 (2020). https://doi.org/10.1016/j.est.2019.101038
53. S.K. Tiwari, A.K. Thakur, A. De Adhikari, Y. Zhu, N. Wang, Current research of graphene-based nanocomposites and their application for supercapacitors. Nano **10**(10), 2046 (2020). https://doi.org/10.3390/nano10102046
54. A. Mendhe, H.S. Panda, A review on electrolytes for supercapacitor device. Discov. Mater. **3**(1), 29 (2023). https://doi.org/10.1007/s43939-023-00065-3
55. M.B. Arvas, H. Gürsu, M. Gencten, Y. Sahin, Preparation of different heteroatom doped graphene oxide based electrodes by electrochemical method and their supercapacitor applications. J Energy Storage **35**, 102328 (2021). https://doi.org/10.1016/j.est.2021.102328
56. Y.Y. Khine, X. Wen, X. Jin, T. Foller, R. Joshi, Functional groups in graphene oxide. Phys. Chem. Chem. Phys. **24**(43), 26337–26355 (2022). https://doi.org/10.1039/D2CP04082D
57. W. Lu, L. Yan, W. Ye, J. Ning, Y. Zhong, Y. Hu, Defect engineering of electrode materials towards superior reaction kinetics for high-performance supercapacitors. J Mater Chem A Mater **10**(29), 15267–15296 (2022). https://doi.org/10.1039/D2TA02930H
58. J. Zhao, A.F. Burke, Electrochemical capacitors: Materials, technologies and performance. Energy Storage Mater **36**, 31–55 (2021). https://doi.org/10.1016/j.ensm.2020.12.013
59. V. Šedajová, P. Jakubec, A. Bakandritsos, V. Ranc, M. Otyepka, New limits for stability of supercapacitor electrode material based on graphene derivative. Nano **10**(9), 1731 (2020). https://doi.org/10.3390/nano10091731
60. J. Affi et al., Electrochemical and capacitive behavior of reduced graphene oxide from green reduction of graphene oxide by urea for supercapacitor electrodes. J. Mater. Sci. Mater. Electron. **34**(22), 1638 (2023). https://doi.org/10.1007/s10854-023-11076-4
61. C. Liu, Z. Yu, D. Neff, A. Zhamu, B.Z. Jang, Graphene-based supercapacitor with an ultrahigh energy density. Nano Lett. **10**(12), 4863–4868 (2010). https://doi.org/10.1021/nl102661q
62. P. Saikia, K. Dutta, A.K. Guha, S.K. Dolui, P. Barman, L.J. Borthakur, High-performance aqueous electrolyte based supercapacitor of carboxylic acid functionalized carbon-nanotubes and graphene nano composite. Mater. Chem. Phys. **258**, 123786 (2021). https://doi.org/10.1016/j.matchemphys.2020.123786
63. L. Cui, Y. Li, M. Jia, C. Cheng, X. Jin, A self-assembled and flexible supercapacitor based on redox-active lignin-based nitrogen-doped activated carbon functionalized graphene hydrogels. J. Electrochem. Soc. **168**(5), 053504 (2021). https://doi.org/10.1149/1945-7111/ac00f6
64. Z. Cai, C. Jiang, X.F. Xiao, Y.S. Zhang, L. Liang, Lignin-based biochar/graphene oxide composites as supercapacitor electrode materials. IOP Conf Ser Mater Sci Eng **359**, 012046 (2018). https://doi.org/10.1088/1757-899X/359/1/012046
65. M. Fu et al., Facile synthesis of V2O5/graphene composites as advanced electrode materials in supercapacitors. J. Alloys Compd. **862**, 158006 (2021). https://doi.org/10.1016/j.jallcom.2020.158006
66. Y. Chai, Z. Li, J. Wang, Z. Mo, S. Yang, Construction of hierarchical holey graphene/MnO2 composites as potential electrode materials for supercapacitors. J. Alloys Compd. **775**, 1206–1212 (2019). https://doi.org/10.1016/j.jallcom.2018.10.259
67. H. Yu et al., Facile fabrication and energy storage analysis of graphene/PANI paper electrodes for supercapacitor application. Electrochim. Acta **253**, 239–247 (2017). https://doi.org/10.1016/j.electacta.2017.09.071
68. M. Hassan et al., Hierarchical assembly of graphene/polyaniline nanostructures to synthesize free-standing supercapacitor electrode. Compos. Sci. Technol. **98**, 1–8 (2014). https://doi.org/10.1016/j.compscitech.2014.04.007

69. L. Zhang et al., Three-dimensional structures of graphene/polyaniline hybrid films constructed by steamed water for high-performance supercapacitors. J. Power Sources **342**, 1–8 (2017). https://doi.org/10.1016/j.jpowsour.2016.11.068
70. A. Moyseowicz, G. Gryglewicz, Hydrothermal-assisted synthesis of a porous polyaniline/reduced graphene oxide composite as a high-performance electrode material for supercapacitors. Compos. Part B Eng. **159**, 4–12 (2019). https://doi.org/10.1016/j.compositesb.2018.09.069
71. J. Zhu et al., Three-dimensional N-doped graphene/polyaniline composite foam for high performance supercapacitors. Appl. Surf. Sci. **428**, 348–355 (2018). https://doi.org/10.1016/j.apsusc.2017.09.148
72. M. Yu et al., Dual-doped molybdenum trioxide nanowires: A bifunctional anode for fiber-shaped asymmetric supercapacitors and microbial fuel cells. Angew. Chem. **128**(23), 6874–6878 (2016). https://doi.org/10.1002/ange.201602631
73. H. Qi et al., Hierarchical nanocarbon-MnO2 electrodes for enhanced electrochemical capacitor performance. Energy Storage Mater **16**, 607–618 (2019). https://doi.org/10.1016/j.ensm.2018.07.019
74. X. Sun, J. Sun, L. Guo, L. Hou, C. Yuan, Understanding the crystal structure-dependent electrochemical capacitance of spinel and rock-salt Ni–Co oxides via density function theory calculations. RSC Adv. **10**(59), 35611–35618 (2020). https://doi.org/10.1039/D0RA05578F
75. R.B. Pujari, A.C. Lokhande, A.R. Shelke, J.H. Kim, C.D. Lokhande, Chemically deposited nano grain composed MoS2 thin films for supercapacitor application. J. Colloid Interface Sci. **496**, 1–7 (2017). https://doi.org/10.1016/j.jcis.2016.11.026
76. X. Liu, P. Carvalho, M.N. Getz, T. Norby, A. Chatzitakis, Black anatase TiO2 nanotubes with tunable orientation for high performance supercapacitors. J. Phys. Chem. C **123**(36), 21931–21940 (2019). https://doi.org/10.1021/acs.jpcc.9b05070
77. S. Yang, X. Song, P. Zhang, L. Gao, Heating-rate-induced porous α-Fe2O3 with controllable pore size and crystallinity grown on graphene for supercapacitors. ACS Appl. Mater. Interfaces **7**(1), 75–79 (2015). https://doi.org/10.1021/am507910f
78. J.W. Lee, T. Kshetri, K.R. Park, N.H. Kim, O.-K. Park, J.H. Lee, Covalently bonded boron nitride quantum dot and reduced graphene oxide composite electrode for highly efficient supercapacitors. Compos. Part B Eng. **222**, 109089 (2021). https://doi.org/10.1016/j.compositesb.2021.109089
79. Z. Xu et al., Investigation on the role of different conductive polymers in supercapacitors based on a zinc sulfide/reduced graphene oxide/conductive polymer ternary composite electrode. RSC Adv. **10**(6), 3122–3129 (2020). https://doi.org/10.1039/C9RA07842H
80. M. Yun et al., Holey graphene interpenetrating networks for boosting high-capacitive $Co3O_4$ electrodes via an electrophoretic deposition process. Ceram. Int. **47**(19), 27210–27216 (2021). https://doi.org/10.1016/j.ceramint.2021.06.143
81. Y. Chen et al., Porous 3D graphene aerogel co-doped with nitrogen and sulfur for high-performance supercapacitors. Nanotechnology **32**(19), 195405 (2021). https://doi.org/10.1088/1361-6528/abdf8d
82. L. Chi et al., 1.6 V high-voltage aqueous symmetric micro-pseudocapacitors based on two-dimensional polypyrrole/graphene nanosheets. Carbon N Y **194**, 240–247 (2022). https://doi.org/10.1016/j.carbon.2022.03.057
83. Q. Liu et al., Ultrathin MoS2 nanosheets hybridizing with reduced graphene oxide for high-performance pseudocapacitors. FlatChem **26**, 100212 (2021). https://doi.org/10.1016/j.flatc.2020.100212
84. M. Eskandari, R. Malekfar, D. Buceta, P. Taboada, NiCo2O4-based nanostructured composites for high-performance pseudocapacitor electrodes. Colloids Surf. A Physicochem. Eng. Asp. **584**, 124039 (2020). https://doi.org/10.1016/j.colsurfa.2019.124039
85. A. Sankar, S.V. Chitra, M. Jayashree, M. Parthibavarman, T. Amirthavarshini, NiO nanoparticles/graphene nanocomposite as high-performance pseudocapacitor electrodes: Design and implementation. Diam. Relat. Mater. **122**, 108804 (2022). https://doi.org/10.1016/j.diamond.2021.108804

86. K. Makgopa, M.S. Ratsoma, K. Raju, L.F. Mabena, K.D. Modibane, One-step hydrothermal synthesis of nitrogen-doped reduced graphene oxide/hausmannite manganese oxide for symmetric and asymmetric pseudocapacitors. ACS Omega **6**(47), 31421–31434 (2021). https://doi.org/10.1021/acsomega.1c02302
87. P. Mandal, S. Bala, S. Poddar, S. Sarkar, H. Shankar Biswas, Fabrication of graphene-Fe_3O_4-Polypyrrole based ternary material as an electrode for Pseudocapacitor application. Mater Today Proc **65**, 1001–1010 (2022). https://doi.org/10.1016/j.matpr.2022.04.103
88. U.V. Gopi, P. Bhojane, K.S. Smaran, Engineering oxygen-deficient nanocomposite comprising LaNiO3-δ and reduced graphene oxide for high-performance pseudocapacitors. J Energy Storage **54**, 105301 (2022). https://doi.org/10.1016/j.est.2022.105301
89. A. Khazaeli, G. Godbille-Cardona, D.P.J. Barz, A novel flexible hybrid battery–supercapacitor based on a self-assembled vanadium-graphene hydrogel. Adv. Funct. Mater. **30**(21) (2020). https://doi.org/10.1002/adfm.201910738
90. Z. Sun, F. Li, Z. Ma, Q. Wang, F. Qu, Battery-type phosphorus doped FeS2 grown on graphene as anode for hybrid supercapacitor with enhanced specific capacity. J. Alloys Compd. **854**, 157114 (2021). https://doi.org/10.1016/j.jallcom.2020.157114
91. H. Xing et al., Ultrathin and highly crumpled/porous CoP nanosheet arrays anchored on graphene boosts the capacitance and their synergistic effect toward high-performance battery-type hybrid supercapacitors. ACS Appl. Mater. Interfaces **13**(22), 26373–26383 (2021). https://doi.org/10.1021/acsami.1c04921
92. J.R. Rani, R. Thangavel, M. Kim, Y.S. Lee, J.-H. Jang, Ultra-high energy density hybrid supercapacitors using MnO2/reduced graphene oxide hybrid nanoscrolls. Nano **10**(10), 2049 (2020). https://doi.org/10.3390/nano10102049
93. T.D. Le et al., Green flexible graphene–inorganic-hybrid micro-supercapacitors made of fallen leaves enabled by ultrafast laser pulses. Adv. Funct. Mater. **32**(20) (2022). https://doi.org/10.1002/adfm.202107768
94. R. Sahoo, T.H. Lee, D.T. Pham, T.H.T. Luu, Y.H. Lee, Fast-charging high-energy battery–supercapacitor hybrid: anodic reduced graphene oxide–vanadium(iv) oxide sheet-on-sheet heterostructure. ACS Nano **13**(9), 10776–10786 (2019). https://doi.org/10.1021/acsnano.9b05605

Chapter 3
Graphene-Powered Batteries for the Future

3.1 Introduction

Graphene has received widespread recognition as a two-dimensional material in the past two decades due to its unique honeycomb structure of carbon atoms and incomparable properties. The spacing between the carbon bonds is approximately 0.142 nm [1]. The basic structure of graphene is depicted in Fig. 3.1. The foremost factors for the rising popularity of graphene are due to its vast surface area, thin thickness, lower density, excellent mechanical strength, and conductivity. Varied graphene derivatives such as graphene oxide, graphene nanosheets, reduced graphene oxide, and surface functionalised graphene have also displayed excellent behaviour in harsh conditions. Recent researches indicate the high utilisation of graphene in numerous energy-conserving systems as an electrode material for enhancing the performance of these devices. However, as an individual electrode material, graphene exhibits some limitations, such as the high tendency to restacking, inferior ion diffusion, and reduced capacity [2]. To address the shortcomings of graphene electrodes, different methods have been developed to further enhance the performances of the graphene-supported electrodes in energy-storing devices. Graphene can be incorporated with different materials to form composite electrodes, functionalised to enhance its electrochemical performance, and doped with distinct materials for further enhancing its ability to be employed as a substantial electrode means in batteries.

V. Khanna, C. Prakash, *Graphene for Electrochemical Energy Storage*, Lecture Notes in Nanoscale Science and Technology 24,
https://doi.org/10.1007/978-3-031-87643-1_3

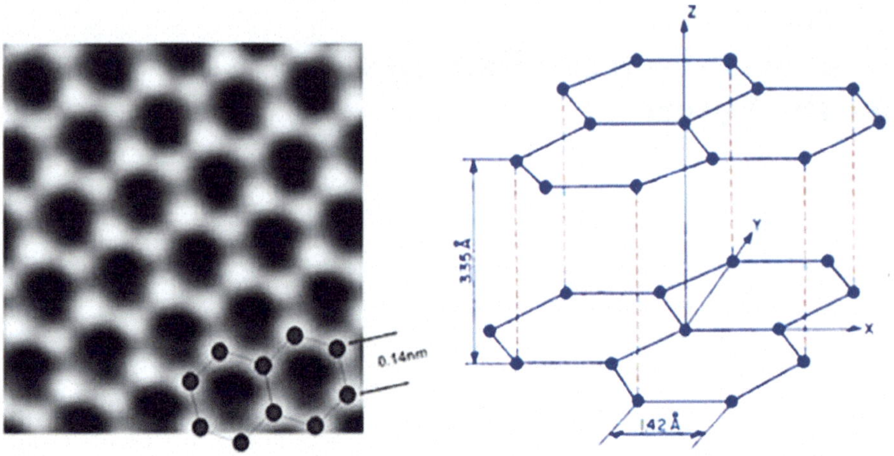

Fig. 3.1 Illustrates the structure of two-dimensional graphene and the spacing between the two layers of graphene [3]

3.2 Revolutionising Batteries

The evolution of new-generation batteries with high capacity, highly efficient performance and prolonged lifespan are vital for devices storing energy. Considering the importance of next-gen batteries, researchers are constantly applying efforts to build electrodes of materials which poses outstanding electrochemical properties, chemically stable and mechanically durable. In this regard, graphene has been identified as the most suitable electrode material owing to its larger surface area, high conductivity as well as mechanical strength, and flexibility to be employed as a cathode as well as anode within batteries and other electrochemical energy storage devices. The application of graphene-based composite material electrodes offers great potential to address the challenges of current battery technologies.

Traditional battery technologies, such as metal-ion batteries, have surpassed other battery systems as they possess a great ability to produce considerably larger energy density, life cycle, and stability. However, due to its incompetence in delivering high charge/discharge rates, capacity begins to fade, and metals begin to oxidise [4]. Also, some metals tend to chemically react with the electrolyte and reduce the efficiency of the battery. Moreover, costlier raw materials such as cobalt may disrupt supply chain stability and adversely affect the affordability of batteries for commercial usage.

Concerning all the above noteworthy details, the focus of the researchers has completely shifted towards the establishment of batteries that would not only outperform conventional batteries but also are sustainable and cost-effective. Therefore, graphene has been observed as the most reliable material to revolutionise the design of batteries, specifically anodes and cathodes.

3.2.1 Graphene Anodes

Graphite has been broadly exploited as an anode material in batteries owing to its affordability, fair conductivity, and remarkable reversibility [5]. However, its limited capacity and safety concerns about metal-ion insertion within the anode structure make it a less viable material for commercial usage. Conversely, graphene allows metal ions to attach on both sides due to the presence of surface functionalities, higher theoretical capacity and defects. Composite materials based on graphene are being greatly examined as battery anodes.

Silicon-graphene composites have been examined extensively to fabricate anodes for metal-ion batteries owing to their great ability to increase cycling performance [6]. Si poses a large capacity of 4200 mAh/g (theoretically), which is approximately ten folds of graphite and abundantly available. Due to this, it has attracted huge interest in the production of anode material [7]. It is worth noting that during cyclic charging and discharging, anode based on Si exhibits large volume fluctuations [8]. The inclusion of graphene into Si increases both the electrical conductivity and the development of stabilised solid electrolytes, along with inhibiting the Si volume changes. Still, it is very challenging to obtain a uniform distribution of silicon on graphene nanosheets due to the difference in their surface chemistry [9]. Different techniques are being developed to produce Si anode with graphene for the improvement in the electrical conductivity as well as decline volume expansion during cyclic changes. Interestingly, analysis for the anode of Si/graphene aerogel composite in lithium-ion battery (LIBs) was carried out by researchers that exhibited excellent performance. After 500 cycles, it showed higher specific capacity of nearly 1100 mAh/g [10]. Similarly, nitrogen-doped graphene aerogel/SiO_2 nanocomposite was prepared utilising the hydrothermal method. For over 80 cycles, 1052 mAh/g capacity was developed at a current density of a 100 mA/g [11].

In a different study to develop a sodium-ion battery, porous Si in an amorphous state was incorporated with rGO for developing anode material, as shown in Fig. 3.2. The composite depicted greater initial capacity (681.6 mAh/g) and improved cyclic stability, as well as increased electrochemical performance (Fig. 3.2h–j). These outcomes show the synergistic effect of combining porous Si with a large surface area of rGO [12]. Similarly, porous Si/graphene composite was architecture to form anode material using a ball milling process using an alkaline solution of etching. The electrode in operation exhibited better reversible capacity (84 mAh/g) compared with the Si electrode (15.7 mAh/g).

Using simple and cost-effective methods of self-assembling and calcination, Si nanoparticles coated with graphene caged in a graphene nanosheet were produced, as shown in Fig. 3.3a rather than an amorphous shell of carbon, a carbon cage can minimise the internally induced stresses by improving the flexibility of a shell. Also, it increases the capacity to adapt the volume alterations or fluctuations of Si in distinct directions. The caging of graphene allowed the graphene sheets to enhance the conductivity substantially due to the interlinked network, as shown in Fig. 3.3e. At 1 A/g and 100 cycles, the discharging capacity of 1616.1 mAh/g and charging

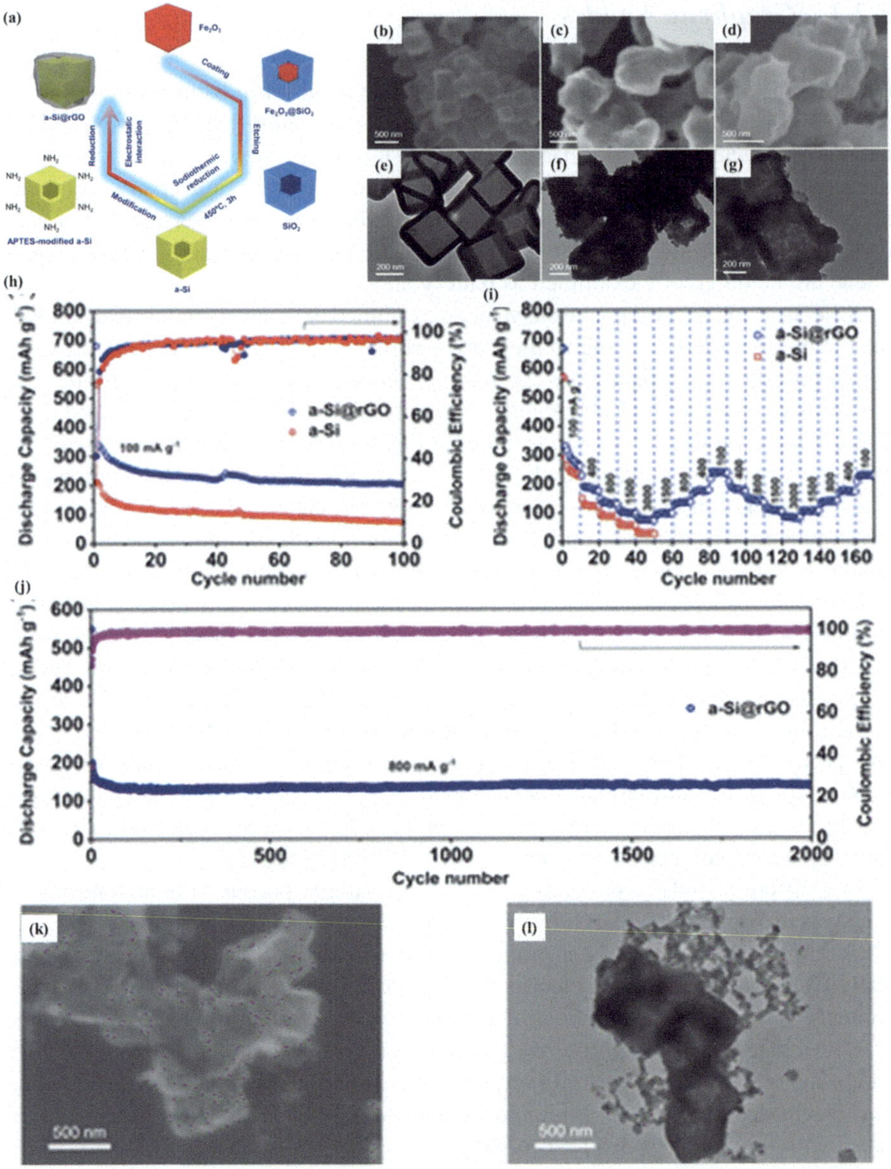

Fig. 3.2 Illustrates (**a**) synthesis of a-Si@rGO composite via etching and reduction method, (**b–d**) SEM images of SiO_2-HNBs, a-Si-HNBs, and a-Si@rGO and their TEM images in (**e–g**), (**h**) compare the electrode material for cyclic stability at 100 mA/g, (**i**) the comparison among the rate capability for different electrode materials, (**j**) cyclic performance of a-Si@rGO electrode at 800 mA/g, at 100–3000 mA/g, and over 50 cycles the (**k**) SEM and (**l**) TEM images of a-Si@rGO electrode [12]

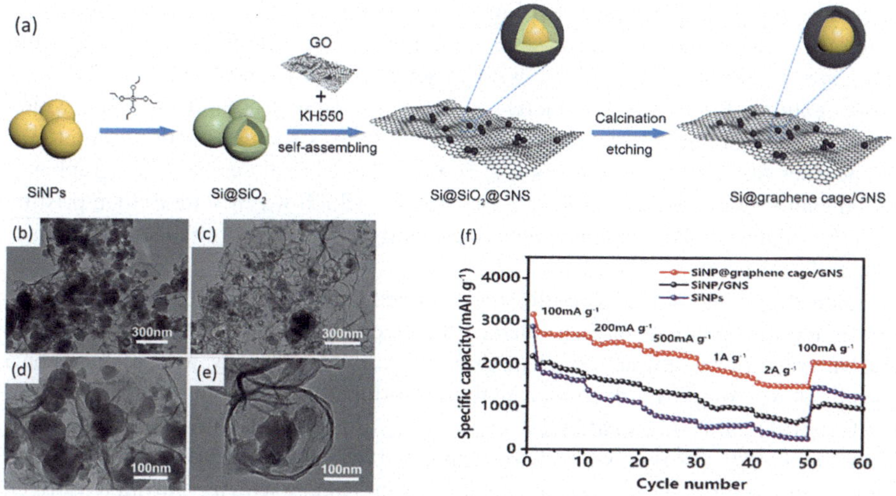

Fig. 3.3 (**a**) fabrication method of Si@graphene cage/GNS, TEM images in low resolution depicting the morphology of (**b**) SiNP/GNS, (**c**) SiNP@graphene caged/GNS, high-resolution TEM images of (**d**) SiNP/GNS, (**e**) SiNP@graphene caged/GNS, and (**f**) comparison of SINPs, SiNP/GNS, and SiNP@graphene cage/GNS in respect with the performance at different levels of current densities [13]

capacity of 1598 mAh/g were delivered [13]. Similarly, two different methods, CVD and thermal reduction, were introduced for preparing the carbon nanocage to control the changes in the volume of Si that occur in the lithiation process. Also, it supports reducing the connectivity in the conductive network. It has resulted in high electrochemical specific capacity in comparison with Si particles. At 5 A/g (current density), it delivered 890 mAh/g of rate capability. Moreover, $LiCoO_2$ as a cathode in a full battery exhibited 329 Wh/kg of energy density [14].

3.2.2 Graphene Cathodes

Graphene has become a highly popular material for batteries largely owing to its extraordinary conductivity, mechanical strength as well as chemical stability. Simple synthesising methods, relatively low-cost production, as well as 170 mAH/g of capacity (theoretically) make olivine lithium ion phosphate ($LiFePO_4$) a considerable cathode material in Li-ion batteries. On the other hand, the minimal electronic conductivity of 10^{-9} S/cm and poor ion diffusion restrict its utility batteries [15]. Therefore, the incorporation of graphene can play a significant role in amplifying its electrochemical ability for batteries. For instance, a composite of $LiFePO_4$ incorporated with very few layers of graphene was prepared through the sol-gel method as a cathode for LiBs. The specific discharge capacity from the galvanostatic discharges at 1C and 2–3.8 voltage for LFP/VFLG increased by ~58%. The cathode composite also depicted a high diffusion rate for lithium ions [16]. Similarly, using different methods such as sonication and drying, $LiFePO_4$ composites were fabricated incorporating graphene with varying content from 0% to 4%. The composite exhibited a highly conductive network in microstructural analysis, which enhanced the ion and electron transfer. The $LiFePO_4$ structure confirmed that graphene does not affect it due to its low content. The specific capacity of graphene-based cathode composite was nearly 142 mAh/g, which was 8% more than pristine $LiFePO_4$ material. The enhancement is evidently due to the high electrical behaviour and surface area of graphene [17].

Interestingly, graphene's ability to be doped with different materials such as boron and nitrogen can further enhance the electrochemical behaviour of cathode for batteries. For example, 162.2 mAh/g of reversible capacity was attained for a metal battery composite decorated with boron-doped graphene at a capacitance of 0.1 C. Further, at 20 C and after 200 cycles, the composite depicted 95.2% retention with 109.9 mAh/g of reversible charge capacity [18]. Similarly, nitrogen-doped graphene nanosheets were incorporated in a Li metal battery with a controlled nitrogen content of 9.26. After 200 cycles and 50 mA/g, the composite exhibited a reversible charge capacity of 344 mAh/g. The capacity reached was 146 mAh/g for over 100 cycles when the current density level was 1 A/g. The doping allows the metal ion to have higher ion diffusion and increased conductivity, which eventually enhances the electrochemical behaviour of the battery cathode [19].

Although the great potential of graphene for metal-based battery cathode has no limits due to its outstanding properties, there exist many challenges that need immediate focus. For example, synthesising graphene in mass production for developing batteries at a large scale, the high cost involved in the production of graphene and its adverse impact on the environment. Apart from this, the wettability of graphene with different materials with high concentrations is another hardship in producing composite cathodes for batteries.

3.3 Graphene in Lithium-Ion Batteries

Lithium-ion batteries (LIBs) have been regarded as an effective and efficient system for energy storage applications as they can provide enhanced density, stability, and reliability. LIBs have found applications in portable, lightweight electronic gadgets, electric vehicles, and power grids [20]. Advancements in the material technology for continuous improvement in the LIBs components, such as separators, electrolytes, anode, and cathode, are highly indispensable (Fig. 3.4). LIBs in China have been predicted to have an energy density of a single cell of 400 and 500 Wh/kg by the end of years 2025 and 2030, respectively. The most primarily focused area for improvement in LIB performance in terms of energy density is electrode materials. Different techniques have evolved, including ball mixing, coating, drying, rolling, vacuum, and filtration to produce a homogenous mixture and reduce material agglomeration. Several parameters, such as concentration of materials, mixing rate,

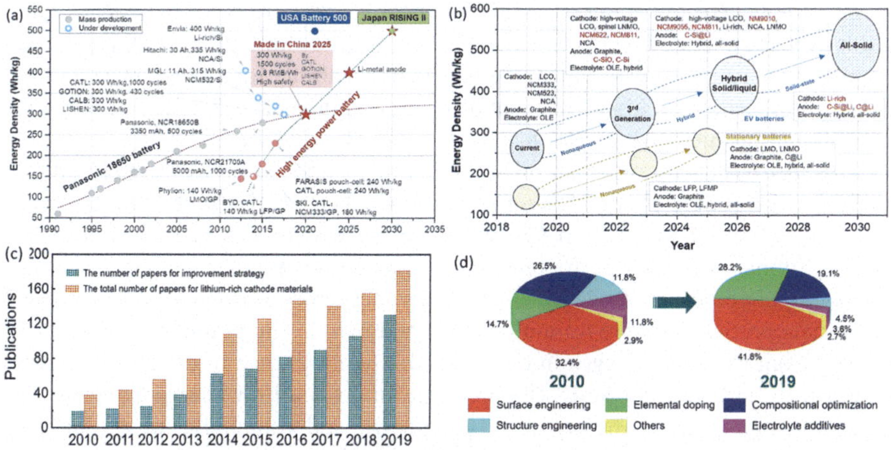

Fig. 3.4 Depicts (**a**) the past, present, and future aspects of LIBs development, (**b**) the route for the development of LIBs in the upcoming period [21], (**c**) the articles published from 2010 to 2019 on LRCMs, and (**d**) comparison of percentage deviations in enhancement strategies in different areas between 2010 and 2019 [22]

and condition, have a critical part to play in defining the strength and conductivity of the manufactured electrodes. Several strategies have been applied in the past decade to enhance the electrochemical behaviour of lithium-rich cathode material composites, which is apparent from the article published in this domain (Fig. 3.4c, d). For further enhancement in their electrochemical behaviour, graphene is incorporated for its high physicochemical properties.

In the initial stages of LIBs researchers, transition metal oxides were highly investigated to enhance the energy and charge storage over repeated cycles. Graphene, due to greater electrochemical characteristics, vast surface area, and adaptability to surface modifications, has now become one of the most reliable materials for energy storage applications in LIBs. For example, Si-doped graphene enhanced the electrode performance extensively and thereby provided high-energy density. Instead of pure graphene, the modified graphene generates larger spacing along with very few SiG defects that contribute to the rapid diffusion of Li ions and electron transfer. Further, the defects at the edge of the graphene surface increased and interacted with the Li ions. Eventually, increasing the capacity of the battery. Also, the bonding between the C-Si-O resulted in controlling the fluctuations in the volume of heteroatom Si particles. Assembling the SiG with $LiFePO_4$ exhibited a high retention rate of 86% after repeated 200 cycles [23].

In an investigation, the SnO_2/RGO composite was created using the one-pot hydrothermal process. The composite demonstrated exceptional strength and a remarkable initial cycle capacity of 1734.1 mAh/g keeping the reversible capacity high (814.7 mAh g − 1) above 70 cycles. Additionally, the fabricated composite exhibits excellent performance at elevated charging rates (596 smAh/g) using 2000 mA/g as the current rate. Furthermore, artificially produced ultrathin silicon nanosheets (Si-NSs) through the etching-reduction technique, when combined with

Li-ion, graphene showed important electrochemical properties for batteries. Higher capacity was generated by the Si-NSs/RGO composites for both the rapid and moderate charge speeds, as well as remarkable cycling stability for more than a thousand cycles. Furthermore, a high rate capability was obtained at 0.05 and 10 Ag − 1, the corresponding values were 2395.8 and 1727.3 mAh/g [24].

The energy density of an Sn incorporated in graphene tubes was analysed in a study. The procedure for the development of the electrode based on graphene composite is depicted in Fig. 3.5a, b. Chemical vapour deposition, along with hydrothermal reactions, was applied to fabricate graphene-integrated electrodes. The Sn particles were confined inside the double tube of graphene. The uniform dispersion of Sn nanoparticles having 40 nm of mean size within the graphene tubes can be viewed from the TEM images in Fig. 3.5c, d. The difference in the energy density obtained by utilising the Sn/DGT composite in comparison with graphite is also evident in Fig. 3.5g, h. The densities (gravimetric and volumetric) of the Sn-DGT electrode after pairing with the Li-based commercial cathode material were observed to be 590 Wh/kg and 1252 Wh/L, respectively [25]. In another study, tin oxide was introduced into the graphene to form an anode electrode for LIBs. A small portion of 5 wt% of graphene content was applied to form an SnO_2/rGO composite electrode. The bond between Sn-O-C was optimised by the structural integrity of SnO_2/rGO formation, delivering high reversible capacity. For 1 A/g, the attained value for over 200 cycles reached 598 mAh/g. In both cases, the volume expansion of tin was

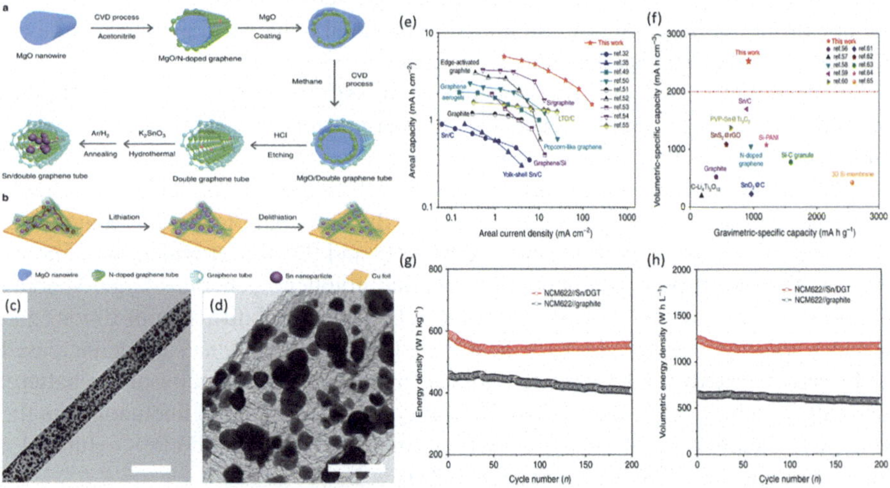

Fig. 3.5 Demonstrates (**a**) procedural steps for fabricating Sn/DGT composite material (**b**) the lithiation and delithiation process (**c**) and (**d**) the morphological images using TEM of Sn/DGT at 500 nm and 100 nm, (**e**) areal performance comparison with different electrode materials with Sn/DGT electrode, (**f**) the high volumetric to gravimetric performance of Sn/DGT electrode compared to other electrodes, (**g**) the energy density of Sn/DGT in NCM622 cell for different cycle numbers, and (**h**) the relation between the volumetric energy density of Sn/DGT in NCM622 cell for different cycle number in comparison to NCM622/graphite electrode [25]

mitigated after modification with carbon, enhancing the interface strengthening of the bonds and resulting in high electrochemical performances [26].

Moving further, nitrogen doping with graphene is another advantageous approach for enhancing the electrochemical behaviour of the batteries. In this regard, using a simple approach of the single-step method, rGO was doped with nitrogen to form NrGO composite electrode for LIBs. Hummers' method has been widely popular and applied to form rGO from GO. NrGO is then obtained by applying thermal annealing using ammonia. At a current density of 10 A/g and after 500 cycles, the reversible capacity obtained with NrGO electrode was 240 mAh/g. Also, the retention capacity was 90%. This has resulted from the ability of pyridinic N to have excellent occupancy within the NrGO [27]. Likewise, $Li_{12}MnO_6Nio_2O_2$ was wrapped in a defective N-doped graphene matrix to generate an excessively conductive electrode to surpass the voltage decay as well as address the rapid capacity fading. At 0.2 C the LMN-NG electrode in comparison to pure LMN electrode generated 6% higher capacity of 286.4 mAh/g. Also, the retention rate for capacity was 86% for LMN-NG after 200 cycling periods. This can be ascribed to the effective grasping of the oxygen group, eventually reducing the decomposition of the electrolyte. Moreover, the microstructural images depicted the high structural stability maintained by the LMN-NG electrode even after prolonged cycling [28].

Lithium doped with boron has been applied in different research to further amplify the electrochemical behaviour of anodes. To illustrate, to limit the growth in the volume of SiO_x@C anodes, Boron dope Li were incorporated with it. While the ion diffusion was enhanced by boron, SiO_x@C permitted high strength to the electrode to resist cracking during cycling periods. After 200 cycles, the attained retention capacity for the SiOx@C-Li/B electrode was 85.4%. The combination of SiOx@C-Li/B and graphite composites can provide a considerable capacity of 531.5 mAh/g in addition to superior cycle stability for LIBs [29]. In a different study, dopamine hydrochloride self-polymerisation with CVD using acetylene along with an amended ball milling process was used for the synthesis process. The flexible innermost carbon layer regulated the volume fluctuations of Si during the process of lithiation and enhanced the electron transport among particles; the boron-doped porous silicon structure improves electron and lithium-ion (Li^+) transport. Meanwhile, the structural integrity of the electrode remained intact during the process. High-capacity porous silicon particles, along with the carbon architecture, allow the composite to promote the electrochemical performance concerning cycle stability and ion storage capacity. Due to this, the B-PSi@double C electrode revealed exceptional cyclic stability in addition to an elevated reversible capacity of 1105.8 mAh/g at a current density level of 0.5 A/g after completing 300 cycles [30]. $LiCoO_2$ (LCO) is well-known in the layered material family due to its extensive research history.

Furthermore, it has been evident from research that different materials can be doped into graphene to extend the ability of anodes in LIBs. For example, In $Li_4Ti_5O_{12}$, porous graphene was incorporated after being doped with nitrogen and sulphur. The density functional theory (DFT) analysis confirmed that co-doping N and S heteroatoms enhance the Li-ion adsorption and diffusion within graphene.

The porous-natured graphene also increased the diffusion channel for Li ions by providing a conductive electron pathway. At 80 °C and 100 °C , the LTO/N, S-PG exhibited 130 and 123 mAh/g of rate capability. Besides this, over 400 cycles and a capacitance of 20C, the electrode retained 83.1% of capacity retention [31].

Graphene nanosheets (GNS) have been incorporated into tungsten disulphide (WS2) to form a composite electrode for analysis as a potential material in LIBs. The strong interface interactions among the composite materials were demonstrated by the newly developed composite produced after high-temperature sintering. The outcomes of the study elaborated on the high and consistent capacity attained by the electrode during electrochemical performance measurements. For more than 200 cycles, the capacity of the developed electrode reached 705 mAh/g. The vast surface area and strongest diffusivity of Li ions led to the better electrochemical capability of the electrode for LIBs [32]. In LIBs, the utilisation of Si/Graphene electrodes has been limited due to practical and scalability issues. However, they can be produced by applying effective strategies. To illustrate, Si/graphene anodes were prepared utilising the ball milling strategy, which is not only a simple but also an economical method of synthetisation. After synthesis, it was discovered that the Si nanoparticles were evenly bonded within the layers and on the surface of the graphene. Over 200 cycles and 100 mA/g current rate, the Si-Graphene composite delivered 1516.23 mAh/g of reversible charge capacity when incorporated within the LIBs as anode. Also, it has demonstrated outstanding excellent performance and prolonged cycling stability [33]. A summary of the electrochemical performance of graphene-based material electrodes in LIBs is exhibited in Table 3.1.

Table 3.1 Electrochemical performances of innumerable combinations of electrode materials based on graphene utilised in LIBs

Electrode material composition	Fabrication technique	Specific capacity	Current density	Cycle number	References
Fe_3O_4@PrGO	Redox deposition	2136 mAH/g	500 mA/g	100	[34]
Co_3O_4/rGO	Electrochemical deposition	1400 mAH/g	2000 mA/g	200	[35]
MnO/C@rGO	Solvothermal along with the carbonisation	1536 mAH/g	100 mA/g	100	[36]
Hollow Co(1-x) S/rGO	Hydrothermal	527 mAh/g	2500 mA/g	107	[37]
NiS/N-rGO	Heat treatment	467 mAH/g	0.5 C	1000	[38]
In situ formed MoS_3/rGO	Electrochemical	900 mAh/g	5 C	2500	[39]
G-Si-CNTs	Microwave irradiation	714 mAH/g	1000 mA/g	400	[23]
SnO_2@MOF/rGO	Wet impregnation	450 mAh/g	1000 mA/g	1000	[40]
NiO@ Co_3O_4-GQDs	Solvothermal	1158 mAH/g	100 mA/g	250	[41]
ZnO/HrGO	Etching	942 mAh/g	200 mA/g	400	[42]

3.4 Graphene in Sodium-Ion Batteries

Sodium belongs to the same class of lithium family and has numerous similar attributes. It has been anticipated that sodium-ion batteries may replace lithium-ion batteries promptly because of the abundant availability of its resources at low cost. The larger size of the sodium ion of 1.02 Å does not allow it to associate with the graphite as an electrode material as in LIBs [43, 44]. For sodium-ion batteries, electrode material selection is extremely crucial to ensure that the charge transportation and sodium-ion insertion, as well as extraction, occur efficiently. When SIBs are being charged, the ions are extracted from the positive and moved across the electrolyte to the negative, just like in LIBs. The cations progressively rise from the negative electrode when the battery is under the charging process. At the same time, the electrons shift towards the negative electrode through the added circuit for balancing the charge. The working principle of SIBs is shown in Fig. 3.6.

Different materials have been used to build SIBs. However, among them, the most commonly employed are 2D materials such as phosphene, silicene, and transition metal dihalides. As discussed in the above section, graphene has been the most widely opted material to produce LIBs due to its unique properties. Similarly, in SIBs, graphene can be directly employed as the negative electrode. For example, a multi-hollow graphene oxide was prepared using the partial reduction method at 300 °C to enhance the graphene structure. Meanwhile, the oxygen-based residual groups attached themselves to the graphene surface and acted as a support element. These points acted as inhibitors for the accumulation of π–π bonds in graphene. Thereby increasing the distance between the layers and increasing sodium ions' movement. For more than 3000 complete cycles, the residual capacity increased to

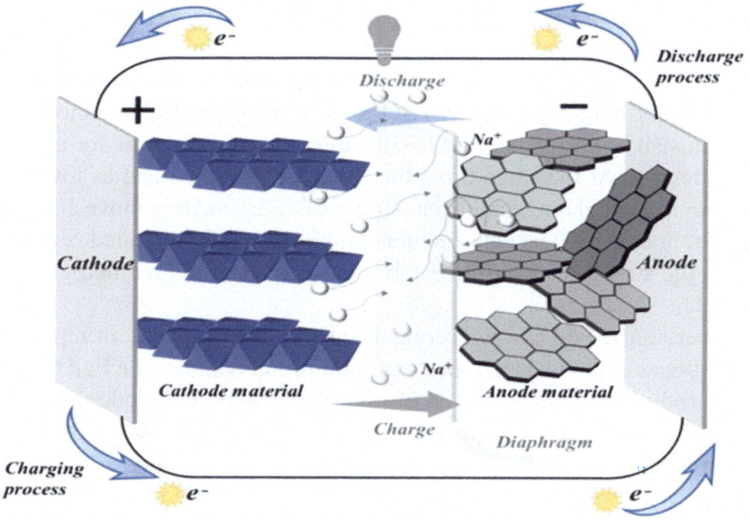

Fig. 3.6 Schematics of a sodium-ion battery mechanism [47]

163 mAh/g at a specific current level of 2 A/g [45]. Further, to control the interlayer distancing, the casting method was employed. The average interlayer distance attained was 0.40 nm with a pore capacity of 0.40 cm^3/g, which is advantageous for sodium ions. Also, the study depicted no structural flaws in the electrode after testing. Over 1000 cycles, the electrode maintained uniformity and exhibited 200 mAh/g of reversible capacity [46].

When cycling procedure is applied, sodium's volume expansion is a critical issue that adversely affects the battery's capacity. Graphene, however, enhances the stability of the electrode. The reversible capacity of $Co_{0.85}Se$-Fe_7Se_8 was 332.3 mAh/g for a 1.0 A/g level of current density over 1000 cycles. The $Co_{0.85}Se$-Fe_7Se_8 interacted with sodium ions and generated several intermediate phases [48]. Similarly, CoS_2 nanoparticles were embedded within the graphene nanoscrolls (GNS) to construct advanced anode electrodes for NIBs. The $CoSe_2$/C@GNS anode at 0.2 A/g has revealed the extended reversible capacity of 545 mAh/g along with rate capabilities as high as 212.5 mAh/g at a specific current of 50 A/g. Also, high cyclic stability was recorded (455 mAh/g) for 5000 cycles and a current density value of 1 A/g [49].

For high-performance NIBs, nitrogen-doped graphene oxide (NDGO) was investigated as a potential anode material. To fully realise the process of the adsorption cycle stability, along with the electrochemical characteristics of NDGO, first-principles simulations were also conducted. It was interesting to find the outcome of the epoxide groups in NDGO affect the adsorption process of sodium ions and the presence of advantageous adsorption sites on the graphene nanosheets. The energy of the Na-ions adsorption was close to the epoxide groups at about −2.38 eV, whereas the adsorption energy of the Na-ions further from the epoxide groups was around −0.45 eV. Nanosheets exhibit metallic behaviour during sedation. For different epoxide groups, namely mono, di, and tri groups with NDGO, the specific capacities attained were 714 mAh/g, 1034 mAh/g, and 1149.32 mAh/g, respectively. This attainment was 2.5–3.5 folds that of hard carbon for sodium-ion batteries [50]. Graphene modified with an amine group served as the anchoring structure for $FeSb_2S_4$ to form the $FeSb_2S_4$/EN-rGO composite electrode. This anode electrode delivered outstanding capacity of 515.7 mAh/g above 500 cycles for a 5 A/g level of current density. At the same time, the capacity loss remained as low as 0.04%/cycle. It also exhibited a reversible capability of 782.5 mAh/g above 100 cycles for 0.1 A/g current density. The enhancement in the functioning of anode composites is ascribed to the strong relationship established between the modified graphene and $FeSb_2S_4$ [51].

In another study, Fe_3O_4 was encompassed into the graphene in replacement to graphite to prepare anodic electrode for a sodium-ion battery. The Fe_3O_4, after coating with carbon incorporated in rGO to form an anodic electrode, had a lamellar structure that exhibited excellent performance in terms of rate performance and cycle stability. Over 300 cycles, considerable discharging capacity (nearly 100 mAh/g) and reversible capability (429 mAh/g) were witnessed when used for SIBs. It was due to a slim layer of carbon that inhibited the growth in volume of Fe_3O_4 and reduced the development of agglomerations [52]. In a different study, a lightweight and highly flexible Fe_3O_4-graphene aerogels composite was prepared by hydrothermal self-assembly technique, as shown in Fig. 3.7a, b. From Fig. 3.7c, it is evident that the rate performance of the Fe_3O_4/rGO anode is higher with 220 mAh/g of initial capacity at 100 mA/g of current density while the reversible was 82 mAh/g for a current density level of 10 A/g [53]. Different combinations of graphene electrode materials for SIBs are depicted in Table 3.2.

The effective fabrication of hybridised porous graphene composite with carbon nanotube and tin oxide (PP-GCS) fibre involved wet spinning, freeze-drying, as well as mechanically powered rearrangement. The mechanically powered structure

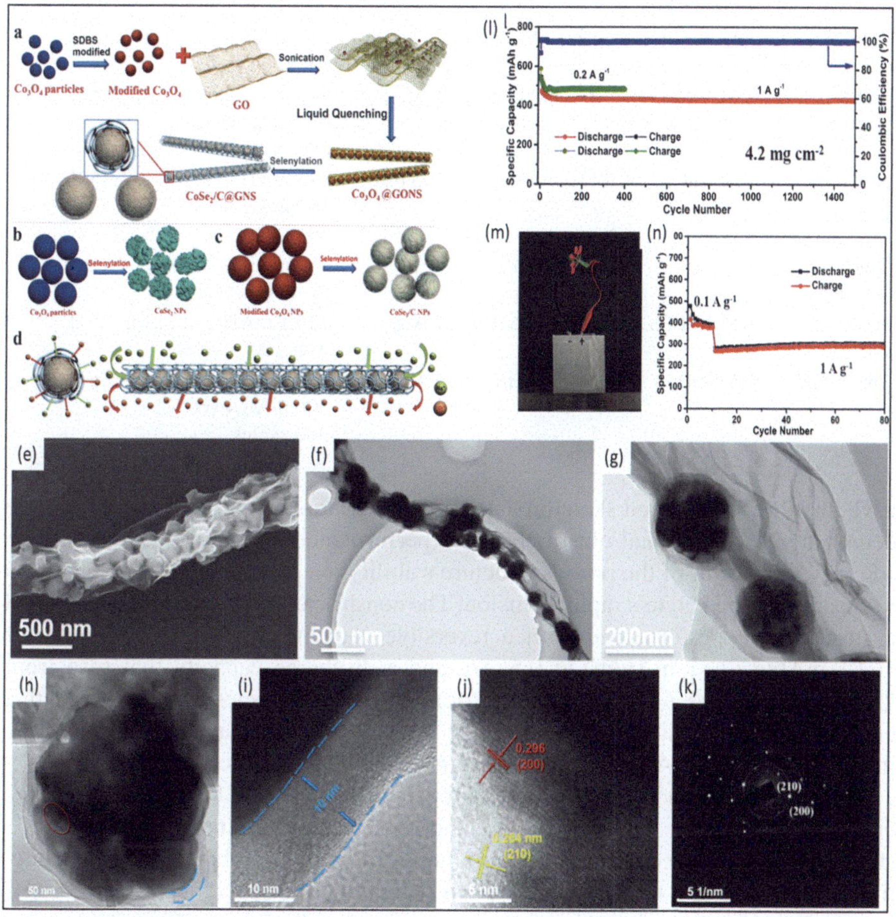

Fig. 3.7 Demonstrates (**a**) the process of developing $CoSe_2$/C/GNS composite after liquid quenching, (**b**) and (**c**) development of $CoSe_2$ NPs and $CoSe_2$/C NPs for comparison with $CoSe_2$/C/GNS, (**d**) developed materials caged into graphene nanoscrolls, (**e**–**k**) high and low magnification microstructural images of developed materials, (**l**) comparison of specific capacity concerning several cycles, (**m**) schematic representation of electrical arrangement, and (**n**) charge and discharge capacity of composite materials against several cycles and different current densities [49]

Table 3.2 Summary of electrochemical performance of graphene-based material for application in SIBs

Electrode structure	Type of electrolyte	Rate capability	Current density	Cyclic performance	References
SnS@Co1-x S-rGO	1 M $NaClO_4$ with 5%FEC	352.92 mAH/g	10 A/g	571.23 mAh/g at 0.1 A/g after 80 cycle	[54]
SnO_2/ C/RGO		105.1 mAH/g	5 A/g	370.3 at 0.2 A/g after 600 cycle	[55]
Fe_3O_4@C/ rGO	1 M $NaCF_3SO_3$	366 mAH/g	0.1 A/g	429 mAH/g at 0.1 A/g over 300 cycle	[52]
Sb/rGO	1 M $NaClO_4$ with 5% FEC	390 mAH/g	2 A/g	584 mAH/g at 0.1 A/g over 120 cycles	[56]
ZnSe@NC/ rGO	1 M $NaClO_4$	195 mAH/g	5 A/g	170 mAH/g at 5 A/g after 500 cycles	[57]
Sb-O-graphene	1 M $NaClO_4$ with 5% FEC	415 mAH/g	2 A/g	460 mAH/g at 0.25 A/g after 200 cycles	[58]
Na-GO	1 M $NaClO_4$	110 mAH/g	200 mA/g	386 mAH/g at 100 mA/g after 60 cycles	[59]
$CuMn_2O_4$/ graphene		145 mAh/g	2 A/g	164 mAH/g at 1 A/g over 300 cycles	[60]
$Na_2Ti_6O_{13}$/ rGO	1 M NaPF6	161.9 mAH/g	2 A/g	120.1 mAH/g at 2 A/g over 1000 cycles	[61]
SnSb-rGO	1 MNaPF6	300 mAH/g	500 mA/g	320 mAH/g at 500 mA/g over 300 cycles	[62]

gave the fibre its required skeleton and gave it a flexible quality. The PP-GCS fibre demonstrated exceptional electrochemical performance, particularly amazing line capacity, as a result of the porous structure's ability to create enough open channels to ensure the electrolyte's rapid diffusion. The constructed fibre-shaped NIB showed a steady cycle performance with a reversible capacity attained significantly of 164.1 mAh/g (2.71 mAh/m) at 0.5 A/g and a discharge capacity that rose from 290.9 to 309 mAh/g (4.8–5.1 mAh/m) for more than 100 cycles at 0.05 A/g. With a current density adjusted to 0.05 A/g, the maximum line capacity of porous fibre increased to 15.74 mAh/m as the diameter of PP-GCS fibre increased [63].

In a unique method, rGO/CNT composite using highly flexible, binder-free carbonaceous paper was fabricated through the electrostatic spray method (Fig. 3.8a–d). Higher active sites for sodium ions were established due to conductive graphene, while the CNT assisted in increasing the space between the interlayers of the graphene to enhance the kinetics of sodium ions. As a result, after cycling at different current rates, the rGO/CNT paper maintains the mean capacity of 101.4 mAh/g after the current density is at 100 mA/g while providing an improved Na + storage capacity. Up to 300 cycles of cycling stabilised an average capacity of 50 mAh/g at a current density of 200 mA/g. The hybrid 3D nanostructures are responsible for the free-standing rGO/CNT high electrochemical

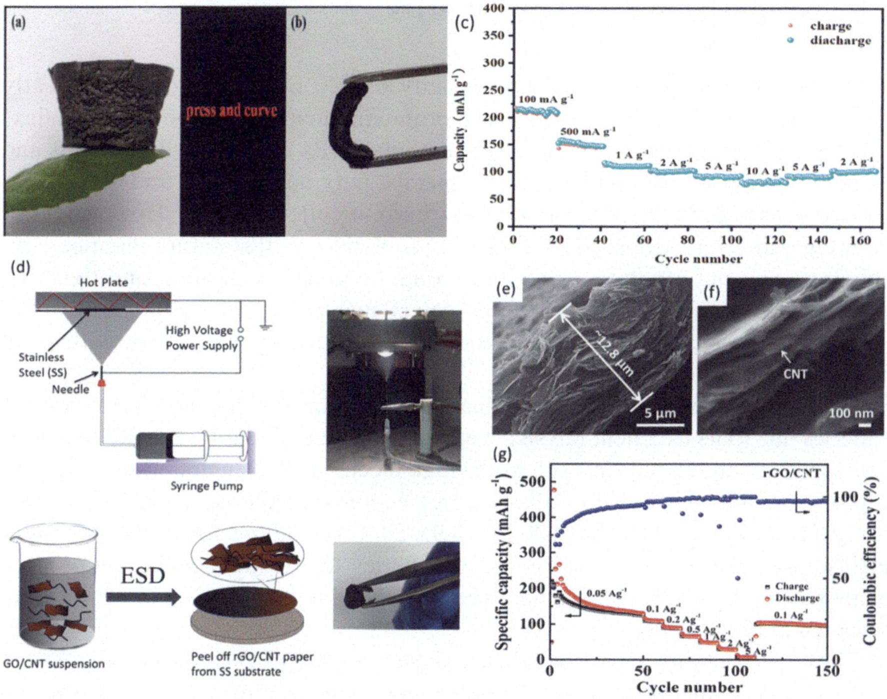

Fig. 3.8 (**a**) Highly compact, lightweight, and (**b**) highly flexible electrode material, (**c**) charge/discharge capacity versus increasing cycle number and current densities [53], (**d**) electrolysis spray method to form binder less paper-based composite for electrode material, (**e**) and (**f**) microsite of developed composite at 5 micrometer and 100 nanometer, and (**g**) displaying the rGO/CNT charge-discharge rate concerning current densities and cycle numbers [64]

performance [64]. Likewise, huge amounts of SeP in an amorphous state were added within the conductive graphene with a crystalline structure. After completing more than 1000 cycles and for a current density of 5 A/g, the composite functioned steadily with 100% of coulomb efficiency [65].

In order to prevent the overlapping process in graphene, SnO_2 was incorporated into the monodispersed graphene via cold quenching and self-assembly methods. The microstructural appearance depicted that the SnO_2 nanoparticles were homogenously dispersed into graphene sheet, restricting its stacking. The outcomes of the study were favourable exhibiting 314 mAh/g of reversible capacity for a 100 mA/g level of current density [66]. Further, in a different research the massive volume changes of $FeMoO_4$ were resolved by adding graphene into it for amplifying the electrochemical behaviour of the composite formed. For 1000 cycles, the electrode attained 145 mAh/g of residual capacity for 0.5 A/g current density. It is evident from the HAADF that the composite electrode is completely in deoxygenated state. The EDS element composition mapping of the composite exhibited even distribution of different elements, namely Fe, Mo, and Na within the electrode material [67].

3.5 Graphene in Zinc-Ion Batteries

Zinc poses remarkable theoretical capacity of 820 mAh/g, highly chemically stable, and lower cost enable ZIBs as a promising material for potable and flexible nano electronics devices. The conventional ZIBs are built with Zn foils and cathodes composed of different active metals, polymer binders, and conductive materials on a metal, in particular copper current collectors [68]. However, poor bonding and interactions, ZIBs confront problem in withstanding the mechanical stress and deformations such as folding, bending, stretching, and twisting. Thus, it becomes highly requisite to select materials which can offer better interface interactions and enable fast electron transports along with high mechanical flexibility [68].

Graphene is a widely opted carbonaceous material for investigating metal-ion batteries due to its excellent physicochemical behaviour and ultrathin 2D structure. The superior conductivity, as well as the high mechanical strength, makes graphene a highly suitable material for utilisation in zinc-based batteries. The different zinc-based battery systems are shown in Fig. 3.9a. However, among these, the most commonly researched is the zinc-ion-based batteries due to their ability to produce better electrochemical performance. The principle diagram of ZIBs in hybrid form is shown in Fig. 3.9b, d.

The cyclic stability and electrochemical performance of ZIBs have been affected adversely due to the formation of dendrites, shape abnormalities, and hydrogen evolution reactions (HER). These factors keep the coulombic efficiency at lower values. To decline the dendritic development and formation, one of the suitable strategies is to include additives within the electrolytes. In general, organic and organic compounds are incorporated into the aqueous ZIBs. Majorly, organic compounds such as polyethene glycol, polyethyleneimine, and diethyl ether are added to the zinc tips patches to limit the localised current density. At the same time, some organic additives, including dimethyl sulfoxide, ethylene glycol, and glucose, have influences on the evaporation phenomenon of solvents for Zn^{2+} ions. To illustrate, 5 mM of vanillin tailored with $ZnSO_4$ electrolyte was prepared for Zn Anode, demonstrating a storing capacity of 10 mAh/cm^2 at 1 mA/cm^2. The coulombic efficiency reached 99.8% with cyclic stability for 1000 h, which was 1 mAh/cm^2. Conversely, in the electrolyte without $ZnSO_4$, the parasitic reactions are triggered due to an unstable Zn anode causing HER corrosion [71].

It is a well-known fact that Mn-based oxides show a high specific capacity of nearly 308 mAh/g. Also, they are abundantly available and inexpensive, and their synthesis is less complex. These aspects of Mn-based oxide materials are appealing for utilisation as cathodes in ZIBs. Nonetheless, the dissolution of Mn3+ ions to Mn2+ ions causes the Jahn-teller effect. Due to the disproportionate dissolution, the ZIBs have a fast reduction of storage capacity. Meanwhile, Mn oxide exhibits poor electron conductivity, causing slow electron transfer. Eventually reducing the rate performance of the ZIBs [72]. By the incorporation of graphene derivatives, these challenges can be reduced to some extent. In a study, flexible ZIBs were created

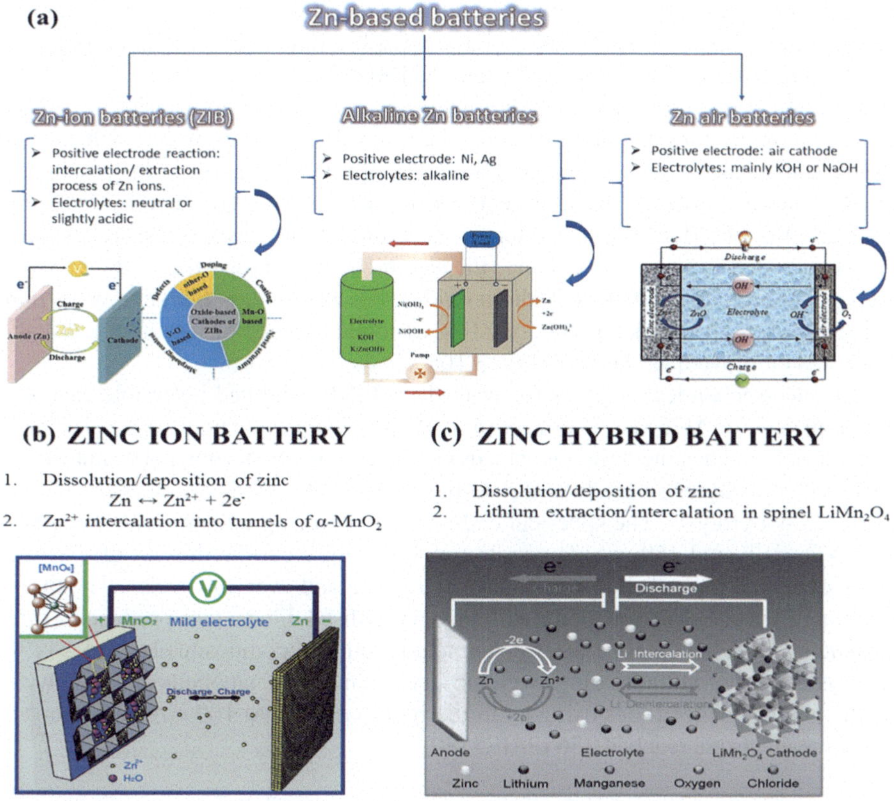

Fig. 3.9 Shows (**a**) various types of batteries available based on zinc [69], the mechanism of (**b**) zinc-ion battery revealing the chemical reaction involved, and (**c**) the intercalation and deintercalation process taking place in the zinc hybrid battery [70]

using MnO_2/rGO using solvothermal as well as vacuum filtration techniques. The membrane made up of 1D MnO_2 nanowires and 2D rGO nanosheets delivered 252 mAh/g of specific capacity [73].

Further, to enhance the cyclic stability of MnO_2 embedded in vertical graphene, the electrode was coated with PEDOT: PSS onto the carbon fibre surface and produced using CVD and hydrothermal treatment in a different study. The prepared composite electrode had a sheet-like structure on the surface of the carbon fibre substrate. The electrode exhibited a capacity retention of nearly 75% for more than 1000 cycles. Also, the electrode had delivered 299 mAh/g of specific capacity for 0.5 A/g rate of current density. The PEDOT: PSS layer established onto the MnO_2/vertical graphene network can reduce the Jahn-Teller distortion effect. This led to the enhancement in the structural integrity as well as extended the cycle life of ZIBs further [74]. Besides this, flower-like spherical nanoparticles of MnO_2 were inserted into the graphene film using thermal reduction techniques. This strategy supported the electrode to generate better electrical conductivity from the graphene film, while the diffusivity of the Zn++ was improved due to the MnO_2 particles in the structure of the composites. Ultimately, the electrode displayed a better specific capacity of 405 mAh/g for over 300 completed cycles at 100 mA/g [75].

Like Mn Oxide materials, there has been separate interest in materials based on vanadium oxides for fabricating cathodes for utilisation in the ZIBs due to their high theoretical capacity and being economically cheaper [76]. Still, they dissolve quickly into electrolytes, causing degradation of their structure, thereby reducing the cycle life of ZIBs [77]. Considering these limitations, several researches have been conducted in the past to enhance the electrochemical properties of the ZIBs by incorporating graphene with vanadium oxides. For example, to fabricate vanadium oxide-based graphene electrodes for ZIBs, a simple strategy of solvothermal reactions was employed, as shown in Fig. 3.10a–c.

The electrochemical analysis for water-based ZIB exhibited reversible capacity reaching 417.7 mAh/g at a specific current of 1.0 A/g for more than 100 cycles [78]. For further enhancement in the electrochemical energy-storing performance of ZIBs with polystyrene-vanadium@rGO-based electrode, open pores were produced using spray pyrolysis. The spherical cathode with porous graphene oxide anchored with VO_2 delivered 159 mAh/g of rate capability for a current density of 0.5 A/g. Also, exhibited high cyclic performance at 1.0 A/g and over 300 cycles. The introduction of porous structure along with tiny crystallised dimensions produces a synergistic effect enhancing the electrochemical kinetics of composite electrodes [79]. In another study, similar findings were observed when VO_2 nanobelts and rGO were made composite electrodes using hydrothermal reactions. At 100 mA/g, the capacity of the electrode reached 456 mAh/g [80].

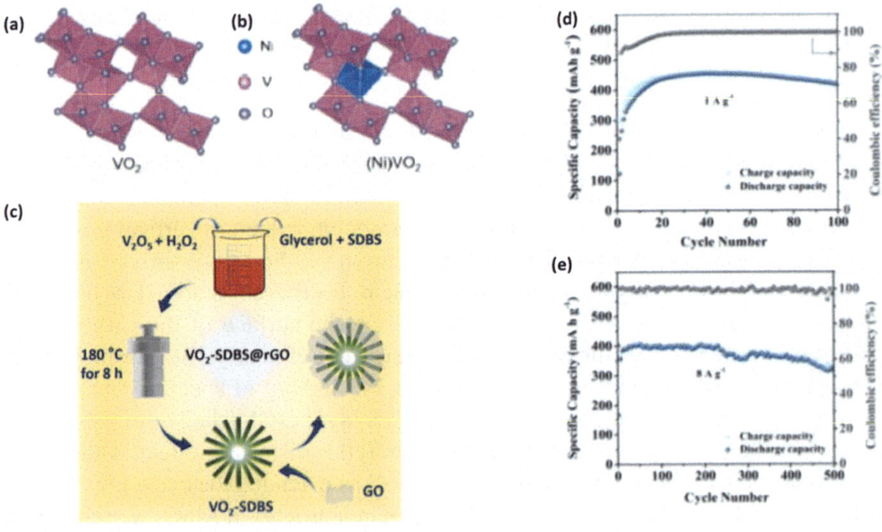

Fig. 3.10 (**a**) The symbolic representation of VO_2 and (**b**) $Ni(VO_2)$, (**c**) the procedure followed to develop VO_2 on rGO, (**d**) and (**e**) show the charge and discharge capacity of the developed electrode for 1 and 8 A/g over a range of cycle numbers [78]

Furthermore, GO sheets wrapped with CuV_2O_6 nanobelts via two different strategies, namely hydrothermal and ultrasonication methods, can effectively address the issue of electrode dissolution in electrolytes. The electrode had 30–46% more specific capacity in comparison with the pure CuV_2O_6. Also, the retention rate for over 3000 cycles was 99.3%, i.e. lowest capacity fading. This is ascribed to the shortest distances available for Zn ions diffusion due to the thin layer of the CuV_2O_6/GO layer [81]. In a different approach, $VO_2 \times 0.26H_2O$ nanobelts were decorated on the rGO as a tendency to work as cathode composite for ZIBs using a simple but effective microwave-assisted hydrothermal technique. The $VO_2 \times 0.26H_2O$@rGO electrode generated 1.000 S/m conductivity, which has been far more compared to pure $VO_2 \times 0.26H_2O$ with 0.324 S/m [82]. Similarly, high electrical conductivity was achieved by VO2/rGO composite electrode produced by hydrothermal method and heat treatment technique for water-based ZIBs. The cathode electrode had a modified layered structure of $Zn_3(OH)_2 V_2O_7 \times 2H_2O$/rGO due to irreversible phase variation enhanced the specific capacity (329.9 mAh/g), with 84% rate of capacity retention for more than 1000 cycles. Moreover, the cathode material exhibited both high power and energy densities of 9796.3 W/kg and 323.3 Wh/kg, respectively [83].

As shown in Fig. 3.11a–f, oxygen-deficient heterojunction in V6O13-x/rGO cathode was developed using the combination of two methods, namely electrostatic assembly and thermal treatment. The plenty of oxygen vacancies within the V6O13-x reduces the interactions with the Zn ions. The transfer rate of Zn ions increases after the electric conductivity is enhanced due to heterointerfaces among the V6O13-x and rGO. The enhanced conductive architecture enabling rapid transfer of metal ions led to improved performance of the electrode, showing higher capacity (424.5 mAh/g) at 0.1 A/g. Also, 96% of retention capacity was attained at the rapid discharge rate of 10 A/g (over 5800 cycles) [84].

Besides metal-oxide materials, other materials have also tempted researchers to enhance the performance of ZIBs. For example, Polyaniline (PANI) is preferred as a cathodic material in hybridisation with graphene for its specific properties, such as conductivity and simple synthesising. Nevertheless, it can degrade the capacity as well as cycle stability of the electrode due to its attribute of deprotonation. However, this issue has been well addressed by researchers by utilising GO in PANI-composed electrodes. It has shown high potential in increasing the storage capacity of Zn ions. The battery with PANI-rGO/CNT electrode exhibited a capacity of 233 mAh/g with a rate performance reaching 100 mAh/g for up to 5 A/g [85]. Similarly, researchers created a PANI/GO composite using an in situ chemical oxidative polymerisation method involving hydrochloric acid along with ammonium persulfate, two oxidants, to decrease the agglomeration of PANI. It was accomplished by exploiting the advantages of GO's inherent chemical properties to entwine with PANI via electrostatic attraction, hydrogen bonding, and π–π interaction. Even after 100 periods of cycles, the structural arrangement of PANI/GO increased the acidic capacity of PANI by two folds to a specific capacity of $\approx$164 mAh/g at a charge/discharge rate of 0.2 C. AT charge/discharge rate of 0.2 C and 95% of retention capacity, the specific capacity of the composite reached to 183 mAh/g after converting GO into rGO with $NaBH_4$ [86].

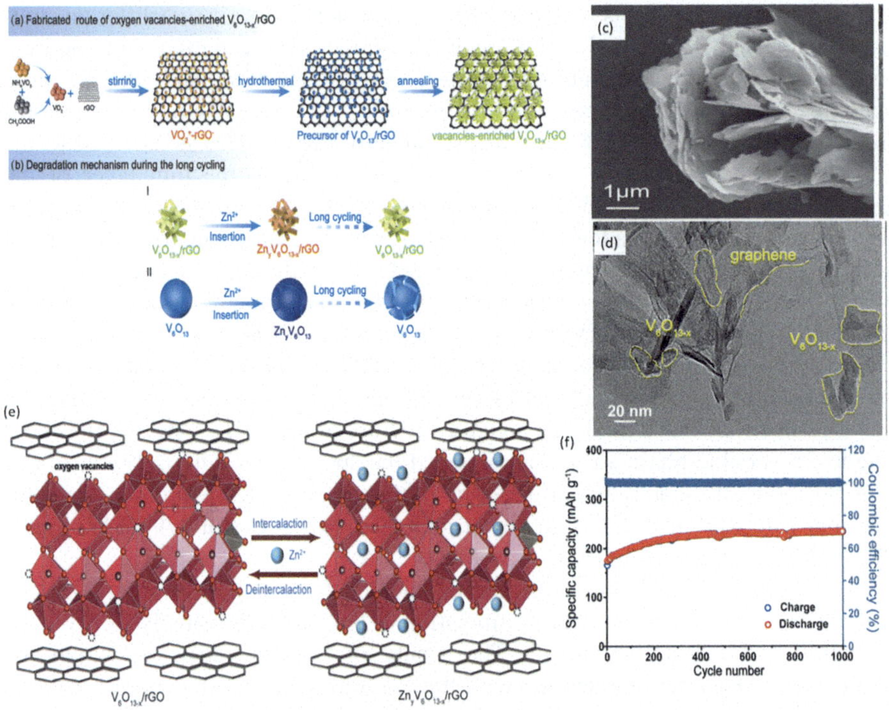

Fig. 3.11 (**a**) the steps involved in producing oxygen vacancies enriched graphene and vanadium oxide composite, (**b**) the process of degrading during prolonged cycles, (**c**) the microstructure of the composite at 1 μm, (**d**) distribution of graphene and vanadium oxide in composite through high magnification image of 1 nm, (**e**) the intercalation and deintercalation of composite with Zn ions, and (**f**) specific capacity of the electrode concerning the number of cycles [84]

3.6 Conclusion

Graphene possesses remarkable electrical conductivity, adjustable space among layers, and a larger surface area and that have turned it into a groundbreaking material that has substantially improved battery technology. Lithium-ion, sodium-ion, and zinc-ion batteries are now some of the most researched battery technologies owing to their simple manufacturing processes, affordability, and the ability to provide significant electrochemical performances. Due to the increased demand for high-performance energy storage solutions, scientific and commercial research is intended to create unique materials combinations that can significantly improve the electrochemical performance of metal-ion batteries. When incorporated into anode and cathode materials, graphene delivered an increased rate of ion diffusion and provided a larger surface area, which is highly essential for traits for metal-ion batteries. Batteries would attain high capacitance and sustain improved performance over prolonged charge-discharge cycles owing to the graphene that improved rate capability, capacity retention, and long-term stability. The development in LIBs,

SIBs, and ZIBs demonstrates the indispensable function that graphene has in enhancing the energy storage capacity of battery technologies efficiently.

References

1. Z. Zhen, H. Zhu, Structure and properties of graphene, in *Graphene*, (Elsevier, 2018), pp. 1–12. https://doi.org/10.1016/B978-0-12-812651-6.00001-X
2. T. Liu, L. Zhang, B. Cheng, X. Hu, J. Yu, Holey graphene for electrochemical energy storage. Cell Rep Phys Sci **1**(10), 100215 (2020). https://doi.org/10.1016/j.xcrp.2020.100215
3. A.G. Olabi, M.A. Abdelkareem, T. Wilberforce, E.T. Sayed, Application of graphene in energy storage device—A review. Renew. Sust. Energ. Rev. **135**, 110026 (2021). https://doi.org/10.1016/j.rser.2020.110026
4. M. Wakihara, Recent developments in lithium ion batteries. Mater Sci Eng R Rep **33**(4), 109–134 (2001). https://doi.org/10.1016/S0927-796X(01)00030-4
5. P. Li, H. Kim, S.-T. Myung, Y.-K. Sun, Diverting exploration of silicon anode into practical way: A review focused on silicon-graphite composite for lithium ion batteries. Energy Storage Mater **35**, 550–576 (2021). https://doi.org/10.1016/j.ensm.2020.11.028
6. H. Torkashvand, M.M. Bagheri-Mohagheghi, Synthesis of Si/rGO nano-composites as anode electrode for lithium-ion battery by CTAB and citrate: Physical properties and voltage–capacity cyclic characterizations. J. Mater. Sci. Mater. Electron. **32**(12), 16456–16466 (2021). https://doi.org/10.1007/s10854-021-06202-z
7. Z. Li et al., Silicon anode with high initial coulombic efficiency by modulated trifunctional binder for high-areal-capacity lithium-ion batteries. Adv. Energy Mater. **10**(20) (2020). https://doi.org/10.1002/aenm.201903110
8. K.T. Kumar et al., Synthesis of graphene-siloxene nanosheet based layered composite materials by tuning its interface chemistry: An efficient anode with overwhelming electrochemical performances for lithium-ion batteries. J. Power Sources **450**, 227618 (2020). https://doi.org/10.1016/j.jpowsour.2019.227618
9. M. Ma, H. Li, Y. Xiong, F. Dong, Rational design, synthesis, and application of silica/graphene-based nanocomposite: A review. Mater. Des. **198**, 109367 (2021). https://doi.org/10.1016/j.matdes.2020.109367
10. F. Tang, T. Jiang, Y. Tan, X. Xu, Y. Zhou, Preparation and electrochemical performance of silicon@graphene aerogel composites for lithium-ion batteries. J. Alloys Compd. **854**, 157135 (2021). https://doi.org/10.1016/j.jallcom.2020.157135
11. X. Dong, X. Zheng, Y. Deng, L. Wang, H. Hong, Z. Ju, SiO2/N-doped graphene aerogel composite anode for lithium-ion batteries. J. Mater. Sci. **55**(27), 13023–13035 (2020). https://doi.org/10.1007/s10853-020-04905-y
12. Y. Zhang et al., Porous amorphous silicon hollow nanoboxes coated with reduced graphene oxide as stable anodes for sodium-ion batteries. ACS Omega **7**(34), 30208–30214 (2022). https://doi.org/10.1021/acsomega.2c03322
13. X.-Y. Han et al., Graphene caging silicon nanoparticles anchored on graphene sheets for high performance Li-ion batteries. Appl. Surf. Sci. **484**, 11–20 (2019). https://doi.org/10.1016/j.apsusc.2019.04.100
14. P. Nie et al., Graphene caging silicon particles for high-performance lithium-ion batteries. Small **14**(25), e1800635 (2018). https://doi.org/10.1002/smll.201800635
15. X. Meng, Recent Progress of graphene as cathode materials in lithium ion batteries. IOP Conf Ser Earth Environ Sci **300**(4), 042039 (2019). https://doi.org/10.1088/1755-1315/300/4/042039
16. A. Amri et al., Novel LiFePO4/very-few-layer-graphene (LFP/VFLG) composites to improve structural and electrochemical properties of lithium-ion battery cathode. Ceram. Int. **50**(11), 19806–19813 (2024). https://doi.org/10.1016/j.ceramint.2024.03.104

17. Y. Fu et al., LiFePO4-graphene composites as high-performance cathodes for lithium-ion batteries: The impact of size and morphology of graphene. Materials **12**(6), 842 (2019). https://doi.org/10.3390/ma12060842
18. C. Li, H. Yuan, Z. Yang, Enhanced electrochemical performance of boron-doped graphene-decorated LiFePO4@C cathode material for lithium-ion batteries. Solid State Ionics **352**, 115366 (2020). https://doi.org/10.1016/j.ssi.2020.115366
19. D. Xiong et al., Superior cathode performance of nitrogen-doped graphene frameworks for lithium ion batteries. ACS Appl. Mater. Interfaces **9**(12), 10643–10651 (2017). https://doi.org/10.1021/acsami.6b15872
20. J.C. Barbosa, R. Gonçalves, C.M. Costa, S. Lanceros-Mendez, Recent advances on materials for lithium-ion batteries. Energies (Basel) **14**(11), 3145 (2021). https://doi.org/10.3390/en14113145
21. Y. Lu, X. Rong, Y.-S. Hu, L. Chen, H. Li, Research and development of advanced battery materials in China. Energy Storage Mater **23**, 144–153 (2019). https://doi.org/10.1016/j.ensm.2019.05.019
22. S. Zhao et al., Towards high-energy-density lithium-ion batteries: Strategies for developing high-capacity lithium-rich cathode materials. Energy Storage Mater **34**, 716–734 (2021). https://doi.org/10.1016/j.ensm.2020.11.008
23. H. Liu et al., Silicon doped graphene as high cycle performance anode for lithium-ion batteries. Carbon N Y **196**, 633–638 (2022). https://doi.org/10.1016/j.carbon.2022.05.018
24. Y. Ren et al., Ultrathin Si nanosheets dispersed in graphene matrix enable stable interface and high rate capability of anode for lithium-ion batteries. Adv Funct Mater **32**(16) (2022). https://doi.org/10.1002/adfm.202110046
25. R. Mo et al., Tin-graphene tubes as anodes for lithium-ion batteries with high volumetric and gravimetric energy densities. Nat. Commun. **11**(1), 1374 (2020). https://doi.org/10.1038/s41467-020-14859-z
26. X. Li et al., Interfacial engineering in SnO2-embedded graphene anode materials for high performance lithium-ion batteries. Sci. Rep. **14**(1), 16751 (2024). https://doi.org/10.1038/s41598-024-67647-w
27. A. Tasdemir, B. Bulut Kopuklu, A.C. Kirlioglu, S. Alkan Gursel, A. Yurum, The influence of nitrogen doping on reduced graphene oxide as highly cyclable Li-ion battery anode with enhanced performance. Int. J. Hydrog. Energy **46**(21), 11865–11877 (2021). https://doi.org/10.1016/j.ijhydene.2021.01.099
28. M. Chen et al., N-doped graphene-modified Li-rich layered Li $_{1.2}$ Mn $_{0.6}$ Ni $_{0.2}$ O $_{2}$ cathode for high-performance Li-ion batteries. ACS Appl Energy Mater **5**(4), 4307–4317 (2022). https://doi.org/10.1021/acsaem.1c03936
29. X.-D. Li et al., Lithium/boron co-doped micrometer SiO_x as promising anode materials for high-energy-density Li-ion batteries. ACS Appl. Mater. Interfaces **14**(24), 27854–27860 (2022). https://doi.org/10.1021/acsami.2c04983
30. Y. Li et al., Multi-functional double carbon shells coated boron-doped porous Si as anode materials for high-performance lithium-ion batteries. Electrochim. Acta **462**, 142712 (2023). https://doi.org/10.1016/j.electacta.2023.142712
31. D. Wang et al., Nitrogen, sulfur co-doped porous graphene boosting Li4Ti5O12 anode performance for high-rate and long-life lithium ion batteries. Energy Storage Mater **27**, 387–395 (2020). https://doi.org/10.1016/j.ensm.2020.02.019
32. Y.-L. Wu, J.-B. Hong, W.-X. Zhong, C.-X. Wang, Z.-F. Li, S. Dmytro, Auxiliary ball milling to prepare WS2/graphene nanosheets composite for lithium-ion battery anode materials. Tungsten **6**(1), 124–133 (2024). https://doi.org/10.1007/s42864-023-00216-2
33. Y. Zhang et al., Functionalization-assistant ball milling towards Si/graphene anodes in high performance Li-ion batteries. Carbon N Y **181**, 300–309 (2021). https://doi.org/10.1016/j.carbon.2021.05.024
34. B. Bulut Kopuklu, A. Tasdemir, S. Alkan Gursel, A. Yurum, High stability graphene oxide aerogel supported ultrafine Fe3O4 particles with superior performance as a Li-ion battery anode. Carbon N Y **174**, 158–172 (2021). https://doi.org/10.1016/j.carbon.2020.12.049

35. K. Wu et al., Hierarchical porous arrays of mesoporous Co3O4 nanosheets grown on graphene skin for high-rate and high-capacity energy storage. J. Alloys Compd. **820**, 153296 (2020). https://doi.org/10.1016/j.jallcom.2019.153296
36. M. Ma, J. Zhang, W. Shen, S. Guo, Cladding transition metal oxide particles with graphene oxide sheets: An efficient protocol to improve their structural stability and lithium ion diffusion rate. J. Solid State Electrochem. **23**(10), 2969–2977 (2019). https://doi.org/10.1007/s10008-019-04390-7
37. S. Lu et al., In situ-formed hollow cobalt sulfide wrapped by reduced graphene oxide as an anode for high-performance Lithium-ion batteries. ACS Appl. Mater. Interfaces **12**(2), 2671–2678 (2020). https://doi.org/10.1021/acsami.9b18931
38. Y.-J. Lee, T.-H. Ha, G.-B. Cho, K.-W. Kim, J.-H. Ahn, K.-K. Cho, Fabrication of nickel sulfide/nitrogen-doped reduced graphene oxide nanocomposite as anode material for lithium-ion batteries and its electrochemical performance. J. Nanosci. Nanotechnol. **20**(11), 6782–6787 (2020). https://doi.org/10.1166/jnn.2020.18783
39. U. Chang et al., In situ self-formed nanosheet MoS3/reduced graphene oxide material showing superior performance as a lithium-ion battery cathode. ACS Nano **13**(2), 1490–1498 (2019). https://doi.org/10.1021/acsnano.8b07191
40. C. Gao, Z. Jiang, P. Wang, L.R. Jensen, Y. Zhang, Y. Yue, Optimized assembling of MOF/SnO2/graphene leads to superior anode for lithium ion batteries. Nano Energy **74**, 104868 (2020). https://doi.org/10.1016/j.nanoen.2020.104868
41. X. Yin, C. Zhi, W. Sun, L.-P. Lv, Y. Wang, Multilayer NiO@Co_3O_4 @graphene quantum dots hollow spheres for high-performance lithium-ion batteries and supercapacitors. J Mater Chem A Mater **7**(13), 7800–7814 (2019). https://doi.org/10.1039/C8TA11982A
42. D. Wu et al., Controllable synthesis of zinc oxide nanoparticles embedded holey reduced graphene oxide nanocomposite as a high-performance anode for lithium-ion batteries. Powder Technol. **367**, 774–781 (2020). https://doi.org/10.1016/j.powtec.2020.04.046
43. Q. Li et al., Advances in metal phosphides for sodium-ion batteries. SusMat **1**(3), 359–392 (2021). https://doi.org/10.1002/sus2.19
44. J. Li et al., Electrochemical Zintl cluster Bi22-induced chemically bonded bismuth/graphene oxide composite for sodium-ion batteries. Electrochim. Acta **413**, 140174 (2022). https://doi.org/10.1016/j.electacta.2022.140174
45. J. Zhao et al., Partially reduced holey graphene oxide as high performance anode for sodium-ion batteries. Adv Energy Mater **9**(7) (2019). https://doi.org/10.1002/aenm.201803215
46. H.-J. Li et al., Nanocasting construction of few-graphene-layers carbon with tunable layer spacing as ultra-stable anode for sodium-ion batteries. Electrochim. Acta **419**, 140423 (2022). https://doi.org/10.1016/j.electacta.2022.140423
47. Q. Jia et al., A review of carbon anode materials for sodium-ion batteries: Key materials, sodium-storage mechanisms, applications, and large-scale design principles. Molecules **29**(18), 4331 (2024). https://doi.org/10.3390/molecules29184331
48. H. Wu et al., Co0.85Se–Fe7Se8 nanocuboids embedded in reduced graphene oxides as cycle-stable anodes for sodium-ion batteries. Carbon N Y **198**, 171–178 (2022). https://doi.org/10.1016/j.carbon.2022.07.022
49. Z. Zhang et al., Exploration of fast ion diffusion kinetics in graphene nanoscrolls encapsulated CoSe2 as advanced anode for high-rate sodium-ion batteries. Carbon N Y **181**, 69–78 (2021). https://doi.org/10.1016/j.carbon.2021.04.095
50. S.U.D. Shamim et al., Understanding Na-ion adsorption in nitrogen doped graphene oxide anode for rechargeable sodium ion batteries. Appl. Surf. Sci. **579**, 152147 (2022). https://doi.org/10.1016/j.apsusc.2021.152147
51. Q. Peng et al., FeSb2S4 anchored on amine-modified graphene towards high-performance anode material for sodium ion batteries. Chem. Eng. J. **385**, 123857 (2020). https://doi.org/10.1016/j.cej.2019.123857
52. J.-L. Xu et al., In-situ plantation of Fe3O4@C nanoparticles on reduced graphene oxide nanosheet as high-performance anode for lithium/sodium-ion batteries. Appl. Surf. Sci. **546**, 149163 (2021). https://doi.org/10.1016/j.apsusc.2021.149163

53. G. Wei, J. Zhou, M. Xie, Fe_3O_4/graphene aerogel as high-performance flexible anode for sodium-ion battery. J. Phys. Conf. Ser. **1637**(1), 012079 (2020). https://doi.org/10.1088/1742-6596/1637/1/012079
54. L. Zhang et al., N-doped Ni-co bimetallic derived hollow nano-framework cubes anchored on 3D reduced graphene aerogel with enhanced sodium ion batteries performance. Colloids Surf. A Physicochem. Eng. Asp. **659**, 130796 (2023). https://doi.org/10.1016/j.colsurfa.2022.130796
55. L. Yu et al., Rational engineering of a carbon skeleton supported tin dioxide nanocomposite from MOF on graphene precursor for superior lithium and sodium ion storage. J. Colloid Interface Sci. **653**, 359–369 (2024). https://doi.org/10.1016/j.jcis.2023.09.065
56. X. Li, J. Qu, H. Xie, Q. Song, G. Fu, H. Yin, An electro-deoxidation approach to co-converting antimony oxide/graphene oxide to antimony/graphene composite for sodium-ion battery anode. Electrochim. Acta **332**, 135501 (2020). https://doi.org/10.1016/j.electacta.2019.135501
57. L. Zhang et al., Rational designed hierarchical dual carbon protected ZnSe anode for advanced sodium-ion hybrid capacitors. J Energy Storage **52**, 104970 (2022). https://doi.org/10.1016/j.est.2022.104970
58. F. Wan et al., In situ binding Sb nanospheres on graphene via oxygen bonds as superior anode for ultrafast sodium-ion batteries. ACS Appl. Mater. Interfaces **8**(12), 7790–7799 (2016). https://doi.org/10.1021/acsami.5b12242
59. M. Kigozi et al., Electrochemical exfoliation and deposition of sodium-graphene oxide composite for high specific capacity cathode/anode for dual-carbon sodium ion battery application. Chem. Phys. Lett. **822**, 140499 (2023). https://doi.org/10.1016/j.cplett.2023.140499
60. M.K.N. Chandran, G.N.S. Babu, M. Sathish, CuMn2O4 anchored on graphene sheets as a high-performance electrodes for sodium-ion batteries. J Energy Storage **65**, 107346 (2023). https://doi.org/10.1016/j.est.2023.107346
61. S. Chandel, J. Zulkifli, J.K. Singh, A.K. Rai, Reduced graphene oxide (rGO) integrated sodium titanate nanocomposite as a high-rate performance anode material for sodium ion batteries. J. Electroanal. Chem. **939**, 117485 (2023). https://doi.org/10.1016/j.jelechem.2023.117485
62. A. Sohan, A. Kumar, T.N. Narayanan, P. Kollu, Tin antimony alloy based reduced graphene oxide composite for fast charging sodium-ion batteries. J Energy Storage **74**, 109312 (2023). https://doi.org/10.1016/j.est.2023.109312
63. Y. Zhang, Z. Bi, Y. Liang, W. Zuo, G. Xu, M. Zhu, Ultrahigh line-capacity and flexible graphene/carbon nanotube/tin oxide fibers as sodium ion battery anodes. Energy Storage Mater **48**, 35–43 (2022). https://doi.org/10.1016/j.ensm.2022.03.002
64. Y. Hao, C. Wang, Free-standing reduced graphene oxide/carbon nanotube paper for flexible sodium-ion battery applications. Molecules **25**(4), 1014 (2020). https://doi.org/10.3390/molecules25041014
65. J. Sang et al., Effective coupling of amorphous selenium phosphide with high-conductivity graphene as resilient high-capacity anode for sodium-ion batteries. Adv. Funct. Mater. **33**(19) (2023). https://doi.org/10.1002/adfm.202211640
66. Y. Li, Y. Zhao, C. Ma, J. Shi, Y. Zhao, Highly monodispersed graphene/SnO2 hybrid nanosheets as bifunctional anode materials of Li-ion and Na-ion batteries. J. Alloys Compd. **821**, 153506 (2020). https://doi.org/10.1016/j.jallcom.2019.153506
67. G.N. Suresh Babu, N.K. Shakkeel, N. Kalaiselvi, FeMoO4 nanorods anchored on graphene sheets as a potential anode for high performance sodium ion batteries. J. Alloys Compd. **877**, 160306 (2021). https://doi.org/10.1016/j.jallcom.2021.160306
68. L. Wu, M. Zhang, W. Xu, Y. Dong, Recent advances in carbon materials for flexible zinc ion batteries. New Carbon Materials **37**(5), 827–851 (2022). https://doi.org/10.1016/S1872-5805(22)60628-0
69. F. Lionetto, N. Arianpouya, B. Bozzini, A. Maffezzoli, M. Nematollahi, C. Mele, Advances in zinc-ion structural batteries. J Energy Storage **84**, 110849 (2024). https://doi.org/10.1016/j.est.2024.110849
70. A.R. Mainar et al., An overview of progress in electrolytes for secondary zinc-air batteries and other storage systems based on zinc. J Energy Storage **15**, 304–328 (2018). https://doi.org/10.1016/j.est.2017.12.004

71. K. Zhao et al., Stabilizing zinc electrodes with a vanillin additive in mild aqueous electrolytes. ACS Appl. Mater. Interfaces **13**(40), 47650–47658 (2021). https://doi.org/10.1021/acsami.1c14407
72. W. Shi, W.S.V. Lee, J. Xue, Recent development of Mn-based oxides as zinc-ion battery cathode. ChemSusChem **14**(7), 1634–1658 (2021). https://doi.org/10.1002/cssc.202002493
73. J. Wang et al., A highly flexible and lightweight MnO_2/graphene membrane for superior zinc-ion batteries. Adv Funct Mater **31**(7) (2021). https://doi.org/10.1002/adfm.202007397
74. J. Chen et al., A vertical graphene enhanced Zn–MnO_2 flexible battery towards wearable electronic devices. J Mater Chem A Mater **9**(1), 575–584 (2021). https://doi.org/10.1039/D0TA08775K
75. S. Kim, G. An, Zinc-ion battery based on heteroatom-doped improved-quality graphene film as a functional current collector. Int. J. Energy Res. **46**(12), 16658–16669 (2022). https://doi.org/10.1002/er.8328
76. J. Ding, H. Gao, D. Ji, K. Zhao, S. Wang, F. Cheng, Vanadium-based cathodes for aqueous zinc-ion batteries: From crystal structures, diffusion channels to storage mechanisms. J Mater Chem A Mater **9**(9), 5258–5275 (2021). https://doi.org/10.1039/D0TA10336E
77. M. Aizudin et al., Recent advancements of graphene-based materials for zinc-based batteries: Beyond lithium-ion batteries. Small **20**(2), e2305217 (2024). https://doi.org/10.1002/smll.202305217
78. J. Ren, C. Yang, B. Zhou, Preparation and electrochemical properties of vanadium-based anode materials for zinc ion battery. J. Phys. Conf. Ser. **2387**(1), 012034 (2022). https://doi.org/10.1088/1742-6596/2387/1/012034
79. J.H. Choi, J.-S. Park, Y.C. Kang, Macroporous vanadium dioxide-reduced graphene oxide microspheres: Cathode material with enhanced electrochemical kinetics for aqueous zinc-ion batteries. Appl. Surf. Sci. **599**, 153890 (2022). https://doi.org/10.1016/j.apsusc.2022.153890
80. F. Cui, J. Zhao, D. Zhang, Y. Fang, F. Hu, K. Zhu, VO2(B) nanobelts and reduced graphene oxides composites as cathode materials for low-cost rechargeable aqueous zinc ion batteries. Chem. Eng. J. **390**, 124118 (2020). https://doi.org/10.1016/j.cej.2020.124118
81. Y. Liu, Q. Li, K. Ma, G. Yang, C. Wang, Graphene oxide wrapped CuV_2O_6 nanobelts as high-capacity and long-life cathode materials of aqueous zinc-ion batteries. ACS Nano **13**(10), 12081–12089 (2019). https://doi.org/10.1021/acsnano.9b06484
82. L. Xie et al., $VO_2{\cdot}0.26H_2O$ nanobelts@reduced graphene oxides as cathode materials for high-performance aqueous zinc ion batteries. Chem. Commun. **58**(99), 13807–13810 (2022). https://doi.org/10.1039/D2CC04431E
83. Y. Li, J. Wang, Z. Tian, F. Lai, T. Liu, G. He, Phase transformation of VO_2/rGO composites as high-voltage cathodes in zinc-ion batteries. Batter Supercaps **6**(3) (2023). https://doi.org/10.1002/batt.202200509
84. C. Zheng et al., Oxygen-vacancy-reinforced vanadium oxide/graphene heterojunction for accelerated zinc storage with long life span. Small **20**(6) (2024). https://doi.org/10.1002/smll.202306275
85. W. Du et al., Rational-design of polyaniline cathode using proton doping strategy by graphene oxide for enhanced aqueous zinc-ion batteries. J. Power Sources **450**, 227716 (2020). https://doi.org/10.1016/j.jpowsour.2020.227716
86. Z. Wang, J.-J. Han, N. Zhang, D.-D. Sun, T. Han, Synthesis of polyaniline/graphene composite and its application in zinc-rechargeable batteries. J. Solid State Electrochem. **23**(12), 3373–3382 (2019). https://doi.org/10.1007/s10008-019-04435-x

Chapter 4
Energising the Future—Graphene in Fuel Cells and Beyond

4.1 The Evolution of Fuel Cell Technology

The first-ever fuel cell, using hydrogen and oxygen as fuel and platinum electrodes, was developed by Sir William Grove in the early-mid nineteenth century. However, the rapid utilisation and development of internal combustion engines have delayed the further progress of fuel cells. The major factors that affected the advancements of fuel cells are the cost, limited resources, and ineffective functioning of electrochemical means.

The continuous increase in the world population has put a significant load on natural resources to fulfil the rising energy demand, which has led to a scarcity of resources and adversely affected the environment. To address this pressing issue, a highly sustainable, larger, better distribution of power is essential. To boost regulated power distribution around the world, the necessity for advanced strategies and sustainable devices has emerged in the past few decades.

In the mid-twentieth century, NASA makes use of fuel cells for the Gemini and Apollo missions to generate clean and reliable power to be used in life support systems and communication networks. This was a groundbreaking achievement that demonstrated the practical feasibility of fuel cells in energy storage and utilisation. Later, in the 1970s, Kordesch's work played a pioneering role in fuel cell technology by creating a hybrid system combining both alkaline fuel cells and lead-acid batteries. In recent years, the constant focus has been applied to developing fuel cells for innumerable applications, including wearable electronic devices, cell phones, and electric vehicles that demand different power outputs.

Principally, a fuel cell operates on the principle of an electrochemical cell that converts the chemical energy of the fuels, usually hydrogen, directly into clean,

V. Khanna, C. Prakash, *Graphene for Electrochemical Energy Storage*, Lecture Notes in Nanoscale Science and Technology 24,
https://doi.org/10.1007/978-3-031-87643-1_4

(a)

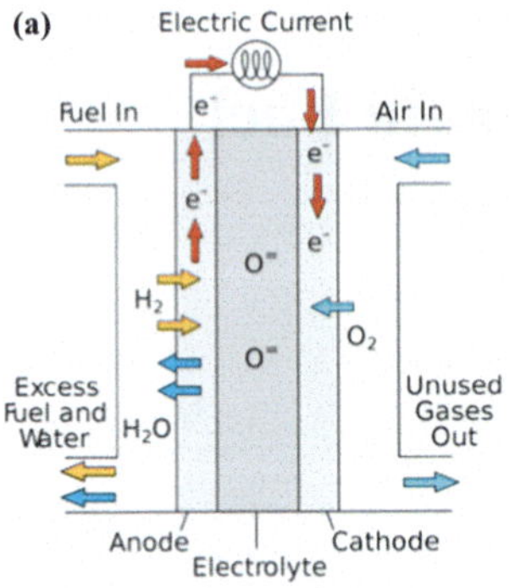

(b)

←Fuel		Anode	Electrolyte		Cathode		Oxidant→
Solid oxide (SOFC)	H_2, CO, CH_4	Ni/ZrO_2	Y-stabilized ZrO_2	←O^{2-}	(La,Sr)MnO_3	O_2 (air)	900 ~1000°C
Molten carbonate (MCFC)	H_2, CO, (CH_4)	Ni based	Alkaline metal carbonate	←CO_3^{2-}	NiO	CO_2+O_2 (CO_2+air)	600 ~700°C
Phosphoric acid (PAFC)	H_2 (CO<1%)	PtRu/C	Conc. phosphoric acid	H^+→	Pt/C	O_2 (air)	160 ~210°C
Polymer electrolyte (PEFC)	H_2 (CO<50 ppm)	Pt/C PtRu/C	Cation exchange membrane	H^+→	Pt/C	O_2 (air)	60 ~80°C
Alkaline (AFC)	H_2(CO_2free)	Ni, PtPd etc	KOHaq	←OH^-	Ag, Ni etc	O_2 (CO_2 free)	5 ~260°C
Fuel→							←Oxidant

Fig. 4.1 (**a**) schematic representation of fuel cell system [2] (**b**) Principle fuel cell technologies with electrolyte materials and their operating temperature ranges [3]

efficient electrical energy [1]. The electrochemical reactions in a fuel cell are specifically characterised by the Nernst equation as given in Eq. (4.1)

$$CE = Eo - \frac{RT}{nF} InQ \tag{4.1}$$

Fuel cells are a class of galvanic cells, and the f Graphene-free energy change is given by the Gibbs free energy using the cell voltage equation.

$$CI\Delta G = -nF\Delta U \tag{4.2}$$

The anodic chemical reactions are categorised as hydrogen and methanol fuel cell reactions and direct oxidation. On the other hand, in cathodic reactions, oxygen reduction occurs, for which the main source is air for the supply of oxygen.

A fuel cell is majorly composed of an ion-conducting membrane, electrodes for fuel and air to conduct electricity, creating three functional layers as shown in Fig. 4.1a. This type of cell is well-known as membrane-electrode assembly (MEA). The moving ions are conducted by the thick, electrically isolating, gas-tight membrane. The electrodes possess porous morphology that enables the diffusion of the fuel and oxygen, electricity, or water and the electrons to move between the active sites, which usually are available near the membrane. Additionally, to enhance the electrochemical interactions within the fuel cell, a catalyst is included in the electrodes.

4.2 Fuel Cells Reimagined with Graphene

The fuel cells can be arranged based on the type of membrane utilised, the electrolyte, and their operational temperature range, as shown in Fig. 4.1b. The most widely known proton exchange membrane fuel cells operate at a relatively low-temperature range using a polymer type of membrane that is the reason they are sometimes regarded as polymer membrane type fuel cells (PEFC). The alkaline membrane type of fuel cell (AFC), utilising hydroxide as the conducting electrolyte, is known to function efficiently between the temperature range of 70 °C to 120 °C. For higher temperature conditions ranging between 150 °C and 210 °C, phosphoric acid fuel cells (PAFC) are preferred. In solid oxide type of fuel cells (SOFC) ceramics, their core membrane operates between 500 and 1000 °C to provide superior ionic conductivity, which may also depend on the thickness and material of the electrolyte.

As the exploration for more advanced materials for enhancing the performance of the fuel cell is progressing, graphene has been considered the most sought-after material.

Graphene, a one-atom-thick sheet of carbon, has tempted the researcher's community with its remarkable properties, including exceptional electrical conductivity, high surface area, and excellent mechanical strength. These attributes of graphene are highly essential for improving the various components of the fuel cell systems. In general, graphene enhances the bipolar plates that are critical for enabling the efficient transfer of electrons and reactants.

The most challenging part of the fuel cell is the development of cost-effective and durable materials. Conventionally, graphite-based metallic plates were examined to enhance the fuel cell capability. However, with added weight, lower durability, and manufacturing complexity, they become outdated. At the same time, graphene has shown better results by addressing these problems to a great extent.

Graphene-based bipolar plates have lower weight due to reduced thickness compared to the conventional counterpart materials. This led to a reduction in the overall size and weight of the fuel cell stack. This is a crucial advantage, particularly for portable devices where weight and size are the deciding factors. Also, the excellent conductivity of graphene is a highly desirable property for the efficient transfer of electrons within the fuel cell, which may improve its overall performance and power output. The utilisation of graphene in fuel cells reaches beyond bipolar plates. It has been observed that graphene can be added to different components of the fuel cell, such as catalyst and membrane, to enhance their performance and strength.

By applying graphene, the development of advanced fuel cell designs can be made that may address the principal challenges this technology is currently encountering. To illustrate, the use of graphene can enhance the power density and efficiency of the fuel cells effectively. Moreover, the integration of graphene-based materials into the catalyst and PEMs can further lead to better catalytic processes and improved conductivity of protons. Thereby reducing the degradation of fuel cell components.

It is, therefore, evident that graphene-based materials have shown potential for significantly improving the capacity of the fuel cells for numerous applications ranging from transportation to power generation and supply. It can be added that the integration of graphene-based materials into the fuel cell systems would be a substantial initiative towards a more sustainable energy future.

4.3 Graphene in Proton Exchange Membrane Fuel Cells

PEM are a highly recognised fuel cell for generating clean power technology. It is known to produce efficient and eco-friendly electric power. Yet, their commercialisation at a broad level is hindered due to numerous challenges, such as insufficient strength of the electrode and high cost. PEMFCs were the first to be introduced in a space program named Gemini. This was a 1 KW fuel stack used as a secondary source for power generation. Polystyrene sulfonate was utilised as the membrane material. However, the membrane exhibited very poor stability. PEM FCs have a similar reaction to that of PAFCs due to their acidic electrolyte (3) and (4).

$$\text{Cathode tw}\frac{1}{2}O2 + 2H + 2e- \rightarrow H2OA \tag{4.3}$$

$$\text{Anode}\, twoH2 \rightarrow 2H + +2e- \tag{4.4}$$

In an investigation, graphene oxide (GO) was introduced as an electrolyte to function at 20% humidity conditions inside the PEMFC. The power generated by the PEMFC using GO was 13 $mWcm^{-2}$, which was ~70% more than delivered by the Nafion membrane with a thickness of 150 µm. The reason could be the decline in the conductivity of the Nafion membrane [4]. However, both the in-plane and through-plane conductivities of the Nafion membrane are higher in comparison to GO. Moreover, the through-plane conductivity is of utmost for fuel cell systems, this has been the major reason for poor outcomes of sprayed GO paper (Fig. 4.2a–d) in comparison with the Nafion membrane as observed in a different study (Fig. 4.2e, f) [5]. The major obstacle in utilising an individual independent GO membrane in PEMFCs is the decrease in functionality groups throughout the process of cell discharge. This results in a gradual loss in power generation, specifically for thin membranes. Majorly, three factors are responsible for this reduction. Firstly, the development of carbonyl groups leads to the formation of H_2O and OH-groups. Secondly, there is a constant decline in OH-groups at the edges and surface of the graphene layer during water formation. Lastly, the synthesis of ether as well as epoxy, along with the presence of hydroxyl [6].

The incomparable physicochemical properties of graphene offer a significant benefit as a catalyst-supporting material in PEMFCs. In a study, different catalysts were tested for durability and catalytic performance in PEMFCs. Among the various platinum-carbon catalysts, the porous nitrogen-doped rGO with platinum exhibited excellent results. The Pt/P-NrGO revealed better durability for a longer period in comparison with the Pt/C catalyst available commercially. The power density, potential, and ECSA of these two materials are shown in Fig. 4.3 [7]. However, it is noteworthy that graphene-based materials may show poor limiting current density, perhaps due to the graphene sheets restacking characteristics.

Interestingly, metal foams had been used to replace carbon plates in PEMFCs due to their porous structure and lower weight benefits. However, the degradation of metal foam due to corrosion caused by the high range of temperature and acidic

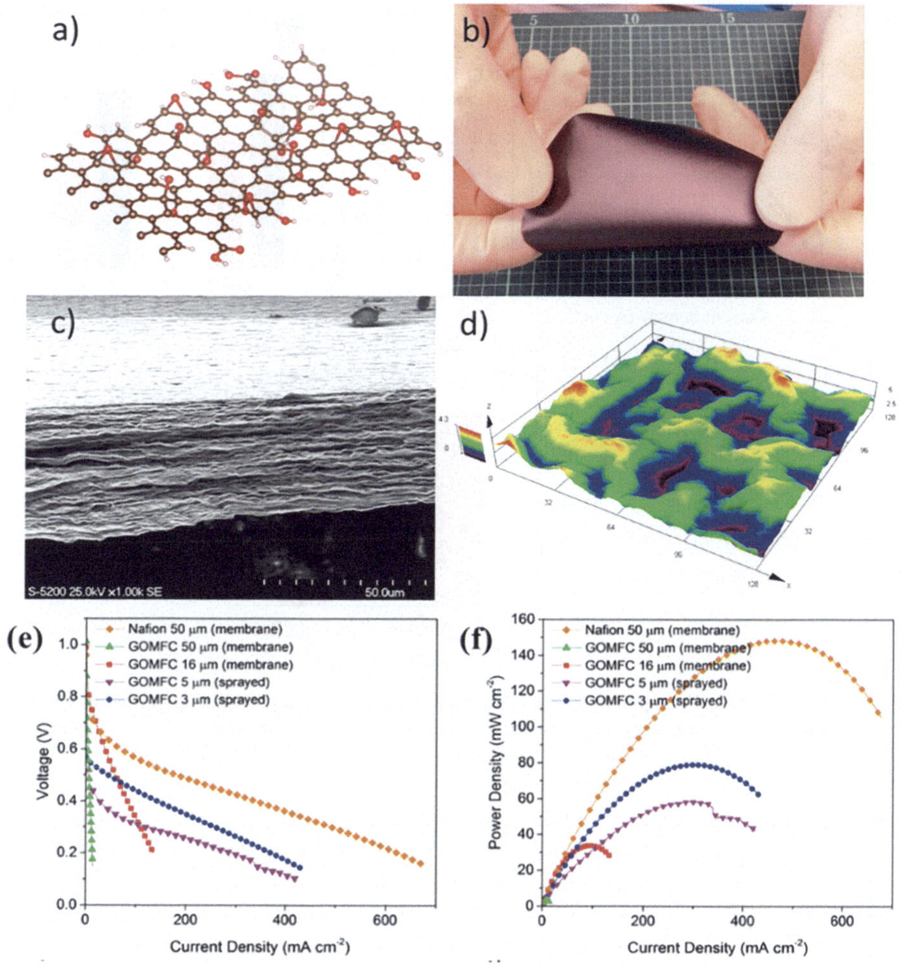

Fig. 4.2 (**a**) represents the structure of the graphene oxide, (**b**) thin graphene oxide paper used in a fuel cell, (**c**) SEM image of the GO paper, (**d**) computational analysis, (**e**) potential voltage of different sample having different sizes, and (**f**) power density of GOMFC and Nafion concerning current density [5]

environment restricted its utilisation in PEMFCs. Addressing this issue, tin-based graphene composite foam was investigated for corrosion. The incorporation of graphene into tin enhanced the corrosion resistance in comparison to tin-based foam. The electrodeposited tin-graphene-coated foam has substantially mitigated the corrosion for a current density applied at 15 $mAcm^{-2}$. It has been observed that in contrast to uncoated foam for temperatures (50 and 80), the corrosion-specific current of Sn/graphene dropped by 41.20% and 38.48%, respectively. Also, the prolonged potential testing for 8 h exhibited a smooth surface, free from any pitting because of the graphene coating. Further, the addition of graphene enabled more Sn

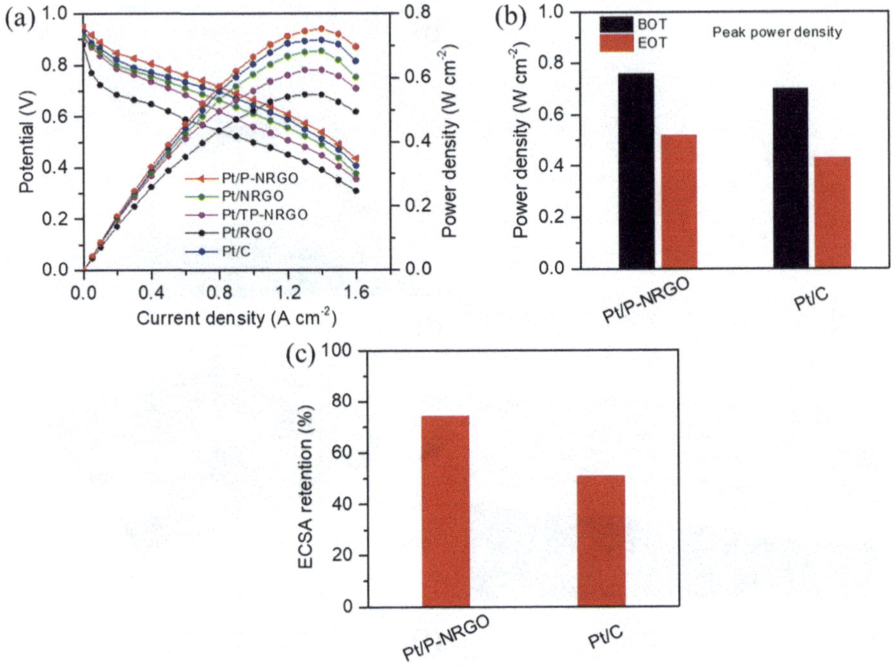

Fig. 4.3 (**a**) Graphical comparison of voltage and power density generated by the GO-based material in a fuel cell concerning current density, compare (**b**) power density and (**c**) ECSA retention of the developed membrane with commercially available membrane for PEMFC [7]

nucleation sites, which reduced the oxidation of Sn, leading to further enhancement in the strength of the composite foam [8]. Similarly, in another corrosion analysis study, Nickel was deposited with graphene to further improve the fuel cell performance. The optimal coating, Ni/G5–10, has shown excellent outcomes in terms of corrosion resistance as well as reduced ohmic resistance, which led to increased fuel cell efficiency (Fig. 4.4a, b) [9].

Further, graphene with 25 wt.% coating on the graphite electrode surface was examined for hydrogen PEMFCs. For performance comparison, an electrode of pristine graphite was also prepared. The coated graphite electrode has been observed to have significantly higher performance through enhanced charge migration. This led to a substantial increase in the cell voltage to 1.0 V with a coated electrode from 0.85 V without a coated electrode. Also, the peak density changed from 53 mAcm^{-2} (pristine graphite) to 79.2 mWcm^{-2} for graphene-coated graphite. The primary factors attributed to the enhancement are reduced ohmic losses and improved charge mobility [10]. Likewise, the surface of the metal-based bipolar plates is coated with graphene to reduce their degradation. To illustrate, stainless steel was utilised as a bipolar plate, which was coated with graphene oxide after reduction. The rGO/316 L SS enabled a reduction in the ion diffusion, leading to the lower potential of 0.155 μAcm^{-2}, which was ~50 times less than the pristine 316 L SS (9.71 μAcm^{-2}) plate [11].

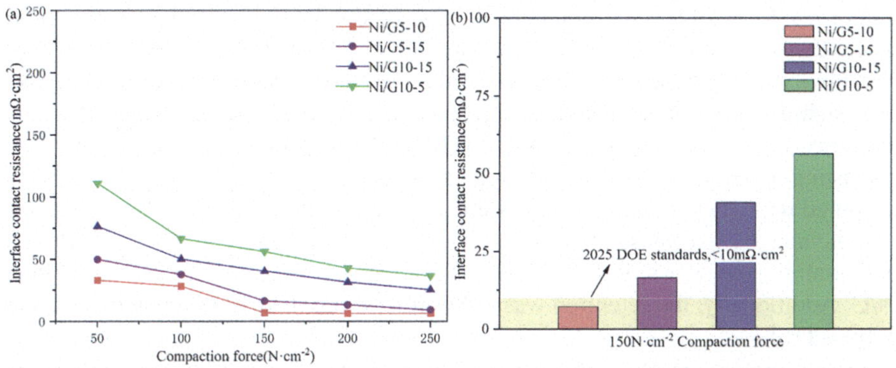

Fig. 4.4 (**a**) Demonstrate the effect of the loading on the resistance at interface contact for different materials and (**b**) increase in the resistance of interface contact for Ni-based graphene membranes for a force of 150 N cm^{-2} [9]

ORR is a deciding factor in fuel cells, particularly at the cathode. The reduction of oxygen molecules to water or hydroxide ions is crucial for optimal performance of fuel cells. Graphene plays a vital role in enhancing the ORR activity within the fuel cell due to its remarkable application as a catalyst and tempting properties. To illustrate, Pt-Pd/S-ZrO_2-GNP were compared with different compositions of pt-Pd added with ZrO_2 and GNP, Pt/S-GNP, Pt-Pd with carbon, and Pt/C electrodes, which exhibited better performance in ORR activity and durability. Also, the ECSA attained was as high as 97.32 m^{2}g^{-1}pt^{-1}. At the same time, the mass activity observed was almost twice that of commercial Pt/C electrodes. Besides, 76% of its original ECSA was retained even after more than 500 cycles. This improvement was a result of improved reaction kinetics due to reduced resistance by the electrode for transfer charges [12]. In a different study, nitrogen derived was doped into rGO derived using the pyrolysis method from the wasted plastic to function as a catalyst within PEMFCs. The findings of the study indicated the notable impact of developed catalysts on the durability and efficiency of the PEMFCs. Later, the catalysts were supported with palladium and ruthenium. The WpNrGO-PD-Ru electrocatalyst achieved a high ECSA (~43 m^{2}g^{-1}) with a potential voltage of 0.33 V for oxidation-reduction reaction (ORR) as well as excellent catalytic performance along with high stability [13].

Furthermore, to efficiently enhance the ORR behaviour in a PEMFC along with the durability and mass activity, platinum was added to the cobalt and hybridised with 3D graphene doped with nitrogen using a simple method of polyol synthesis. In a 0.1 M $HClO_4$, the designed 3D mesoporous graphene-based cathode material demonstrated better ORR in comparison to Pt/C with only a 10 mV of overpotential. The 3D morphology assisted in improving mass transport. This led to the attainment of 1.5 times more mass activity owing to lower Pt. Also, the durability improved for the developed electrode with an ECSA of 5 m^{2}g^{-1} Pt. [14] In a different study, graphene was utilised as a shell for CoPt nanoalloy particles to be worked in PEMFC. The developed catalyst exhibited 681.8 mA mg^{-1} PT of mass activity within 0.1 M $HClO_4$ for a potential voltage of 0.9 V. It also delivers maximum power densities value reaching 2.22 W cm^{-2} for H_2/O_2 and 0.923 W cm^{-2} for H_2/air. The highly strengthened morphology of the developed catalyst exhibited remarkable stability throughout the dynamic state of testing in the condition of lower humidity. Figure 4.5 shows the results of XRD, XPS spectra, ORR, graphene shell consisting CoPt and their morphological structure [15]. Likewise, PtCo were anchored with 3D graphene, which established better ORR. The recorded values for specific activity and mass activity were 1.250 mA cm^{-2} and 0.801 A mg^{-1} Pt at a potential of 0.9 V vs RHE. These measurements were six times more than Pt/C. Additionally, the retention was ~98% after more than 10,000 extensive cycle periods [16].

One intriguing approach that can ease the commercialisation of fuel cells is the development of low-platinum catalysts. Considering this, holey graphene doped with nitrogen was included with platinum particles arranged on the iron to utilise as a catalyst for better ORR. In the computational analysis using density functional theory, it was observed that the adsorption of Pt nanoparticles increased due to the co-doping of iron and nitrogen. Along with this, it also enhanced the adsorption of O_2 with Pt in the ORR [17]. Using the microemulsion technique, Pt-Au-rGO composition was formed to compare with Pt-rGO. The designed graphene-based hybrid materials exhibited increased catalytic behaviour with power density reaching a maximum of 70 mW cm^{-2}. This increment was ~70% of that of Pt-rGO. Further, the

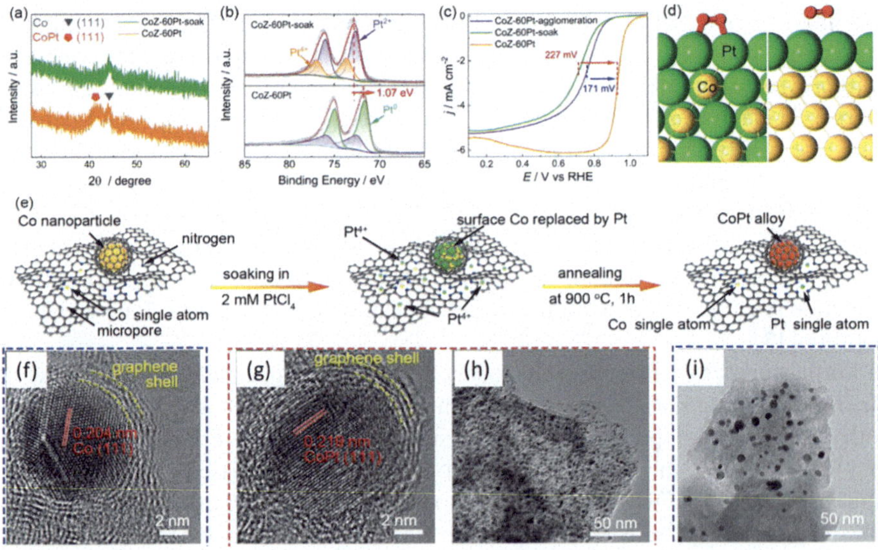

Fig. 4.5 Shows (**a**) the XRD pattern for Co and CoPt materials, (**b**) the binding energy (**c**) ORR in PEMFC, (**d**) the structural appearance of CoPt, € steps involved in obtaining graphene-caged CoPt material, FTIR analysis of (**f**) Co (**g**) CoPt caged in graphene shell, SEM images of (**h**) Co (**i**) CoPt within graphene cage [15]

value increased to 365 mWcm^{-2} during pt-Au arranged with carbon nanotube (CNT) utilised in the FC as a spacer [18].

Interestingly, multiwalled CNTs and graphene were brought together to form a carbonaceous catalyst to examine a high-temperature PEMFC based on a gas-fed polybenzimidazole. The PtRu/MWCNT/rGO catalyst under H_2/air conditions delivered 0.280 W cm^{-2} at 0.180 Acm^{-2} with a potential voltage of 0.6 V. On the other hand, in reformated gas/air conditions, it yielded 0.266 W cm^{-2} for 0.171 A cm^{-2} at 0.6 V [19]. The major reason for the improved electrochemical behaviour of fuel cells using the hybrid rGo catalyst is the synergistic effect produced by the different materials.

Conversely, Pt-free electrocatalysts that are not only effective and durable but also extremely sustainable are essential for improving fuel cell functionality. Considering this, a Co-based ZIF decorated on carbon (C)-doped white graphene (C-WG) was created using a green synthesis technique as an ORR electrocatalyst (Z@C-WG). ZIF and C-WG operate synergistically to improve mass transfer, oxygen adsorption, and active sites. In terms of activity and stability, the hybrid catalyst performs better than Pt/C [20].

4.4 Direct Methanol Fuel Cells

Improving efficiency and performance through graphene, Direct Methanol Fuel Cells (DMFCs) have attracted ample interest because of their high-energy density, straightforward fuel storage, and uncomplicated structure system. Still, their effectiveness is limited by some concerning issues like methanol unwanted diffusion, poor conductivity, and low catalyst durability. Graphene, with its remarkable electrical, mechanical, and chemical properties, could provide radical solutions for DMFCs. In DMFCs, the assembly of materials from covalent organic architectures is crucial for benefiting from the most porous structures. In this regard, $PtFe_3O_4$ was added to graphene aerogels (GAs) for methanol oxidation investigations, and the outcomes were compared with $PtFe_3O_4$ to determine the effects of GA. The inclusion of GAs increased the electrocatalytic functioning by 8.72 times, maintaining 3× higher current after more than 4000 s, and demonstrated high stability for more than 500 cycles. EIS and CO stripping tests confirmed enhanced conductivity and larger surface area [21]. In a similar study, low resistance and effective proton transport to the cathode are ensured by a porous framework of GA. The current density attained by the Pt/50%Nafion/10%GA is 42.62 mW·cm^{-2}, which is 13.5% higher than the conventional type of carbon-catalyst layer (37.56 mW·cm^{-2}). Also, the power density recorded for the GA-doped catalytic layer increased by 95.8% when the temperature was raised from 50 to 70 °C [22]. Further, the introduction of Rh in very small amounts into the MAX/GA catalyst is regarded highly to enhance the DMFC performance. With a large ECSA reaching a maximum of 91.8 m^{2}g^{-1}, superior mass activity attaining 1605.0 mA mg^{-1}, low Tafel slope of 89 mV/decade, decent toxin resistance, and exceptional long-term stability, the Rh/MAX/-rGO catalyst outperforms standards in alkaline methanol oxidation [23].

A well-known energy-efficient technology is the DMFC, in particular when it uses highly conductive materials. For this, forming a composite membrane, modified graphene oxide (GO) along with calcium hydroxy phosphate (CHP) were incorporated into the Nafion membrane. Incorporating 1 wt% CHP and 0.05 wt% tailored GO into Nafion membranes increased the durability by 1.86× and scaled the power density to a maximum of 75.41 mW/cm^2 (1.34× higher than recast Nafion) at 1 M methanol and 70 °C. At 2 M methanol, power density increased to 80.19 mW/cm^2. It dropped significantly to 46.00 mW/cm^2 at 4 M due to extreme methanol crossover. Power densities of CHP/SGO/Nafion concerning current density and time are shown in Fig. 4.6a, b [24]. One-step electrochemical deposition was used to form nickel/graphene/polyaniline (NGP) composites. A constant deposition voltage of 0.5 V was applied for durations ranging from 50 to 250 s. Large Ni particles, around 1–2 μm in size, were observed by scanning electron microscopy embedded in polyaniline-coated graphene sheets. At a peak current density of 5.01 mA cm^{-2}, the NGP composite, which was synthesised at a deposition voltage of 0.5 V for over 200 s, has shown remarkable catalytic efficacy for methanol oxidation. The stability of the composite was ensured by chronoamperometric testing, which shown 12.5% of the original current density was retained for over 1000 s of continuous operation [25].

Further, for fuel cell applications, a scalable and economical process for forming low-layered graphene (LLG) has been developed. While the Raman analysis confirmed the existence of 7–8 layers, the XRD pattern revealed a decline in signal height, signifying the presence of thinner graphene layers. The microstructural analysis depicted the homogenous existence of Pt nanoparticles over carbon support. The developed catalyst performed better and exhibited 57.73 and 21.45 mA cm^{-2} measured values for methanol oxidation [26]. Go sheets have been useful in designing the cation alteration membrane over a grafted cellulose acetate to be used in DMFC. The GCA/GO with 2 wt% exhibited superior selectivity containing 45% water and 0.00599 Scm^{-2}. Also, the permeability for methanol reached 1.12×10^{-7} cm^2 s^{-1} with 26.39×10^3 SS cm^{-3} of electrical selectivity. Moreover, the power density generated by the membrane reached 31.85 mW cm^{-2} for a temperature of 30 °C [27].

Furthermore, Nafion shows poor conductivity as a conventional DMFC membrane therefore, it cannot block methanol. Acid-functionalised modifiers can be used to enhance its capacity since adding poly (vinyl alcohol) (PVA) increases methanol resistance. In a study, a self-healing proton exchange membrane (PEM) with upheld conductivity was developed by combining the sulfonated graphene oxide (SGO) with the Nafion, as well as PVA using a freeze-thaw technique. The hybrid N/PVA/SGO1.0 had a 1.8 times growth in selectivity over recast Nafion due to the combined effect of SGO, which assisted in the enhancement of methanol-blocking by 33.2% and also by increased proton conductivity for 19.5%. Evidently, at 26.48 mA cm^{-2}, the hybrid membrane for 4 M methanol outperformed both Nafion and Nafion 117 by achieving a high-power density (6.07 mW cm^{-2}) at

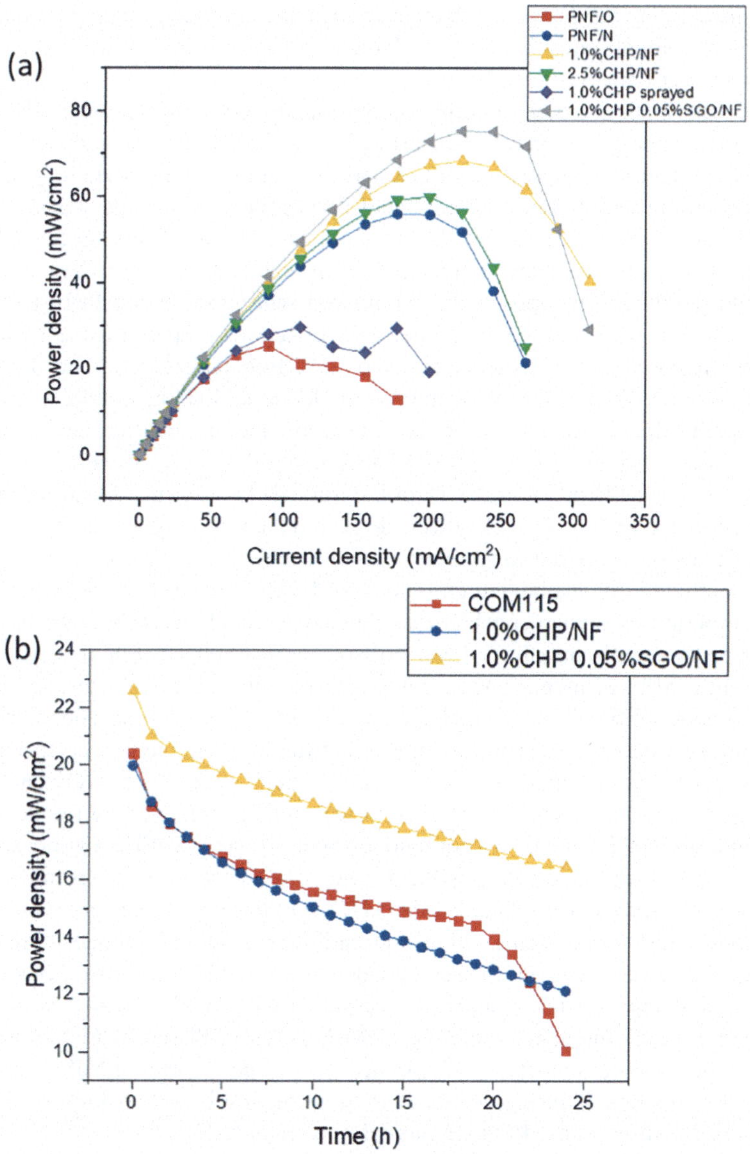

Fig. 4.6 Represents the (**a**) power density of different composite materials versus current density and (**b**) a comparison between the sample of graphene-based material and graphene-free material about power density and time [24]

ambient temperature [28]. In a similar study, to address the poor kinetics and methanol crossover, Gr-coated Nafion PEMs were tested under varying temperatures and methanol concentrations. While at 7.5 M methanol and ambient temperature of 30 °C, Gr reduced methanol crossover by up to 25%, indicating escalated power density from 3.9 to 9.5 mW cm^{-2}. However, the adhesion issues led to Gr delamination at rising temperatures that increased the fuel crossover in contrast to the sample tested for lower temperatures [29].

In an experimental study, titanium oxide was applied to bifunctional PtRu along with graphene aerogel to increase the efficiency of DMFC electrocatalysts. Also, to analyse the effective composition of GA and TiO_2 in the catalysts via response surface methodology. RSM optimisation identified optimal conditions at 202 °C for a Pt to Ru ratio of 1.1 to 1 within 225 of catalyst content, observing 564.87 mA/mgPtRu. Also, the current density reached for the catalysts was 586.15 mAmgPtRu, exhibiting nearly seven times higher compared with commercial PtRu [30]. In a similar research, TiO_2-graphene aerogel with PtRu was prepared using hydrothermal, dry freezing and microwave-assisted reduction techniques. The developed catalyst exhibited about 20% more stability under equal loading conditions in comparison with PtRu/C. Also, the ratio between mass activity and the specific activity obtained was much higher with 608.17 mA mg^{-1}–0.45 mA cm^{-2}PtRu for PtRu/TiO_2-GA than for PtRu/C with 79.11 mAm mg^{-1}–0.19 mA cm^{-2}. Also, in the passive function, the PtRu/TiO_2-GA DMFC delivered 2.6 times higher, i.e. 3.1 mA cm^{-2} of power density compared to commercially available PtRu/C [31].

The commercially available membranes for DMFCs have shown less stability to rising temperatures and are considerably expensive. To address this challenge, polymeric type of membranes have been utilised with graphene. For instance, when fabricating a SPEEK membrane, a composite of both the halloysite clay and the surface functionalised GO were added using the dry phase method. The microstructural analysis confirmed the improvement for the SPEEK membrane in terms of stability during high temperature and hydrophilicity with 2 wt fGO% and 3 wt nanoclay% composite. Additionally, the power density achieved by the developed membrane was 0.47 mS cm^{-1} in comparison with the pure SPEEK membrane with 0.31 mS cm^{-1} [32]. Comparing SPEEK composite proton exchange membranes (PEMs) with sulfonated graphene oxide (SDGO) to those with pure graphene oxide, the former shows better matrix-filler compatibility, leading to bigger hydrophilic domains and fewer vacancies. These enhancements result in improved mechanical qualities, methanol resistance, and proton conductivity. At 80 °C with 1 M methanol and oxygen, direct methanol fuel cells (DMFCs) with SPEEK-SDGO membranes achieve a peak power density of 177.93 mw cm^{-2}, outperforming Nafion 212. The membrane's resilience during accelerated stress testing is further improved by the addition of polydopamine (PDA), a radical scavenger [33].

For the transfer of mass and loading of the catalyst, a layer of diffusion (DL) within the electrode membrane is essential in a microDMFC. Also, the performance of the fuel cell is greatly influenced by the structural design and material of the membrane. As shown in Fig. 4.7a, RGO-based carbon paper was proposed as DL for the anode in μDMFC. The power density achieved with the rGO/CP increased by 53% in contrast to the pure carbon paper diffusion layer. Besides, the developed rGO/CP layer also reduced the resistance to charge transfer effectively [34].

Chitosan-based electrolyte membranes are increasingly developed for fuel cell applications. Chitosan can be utilised with graphene and conductive polymers. For

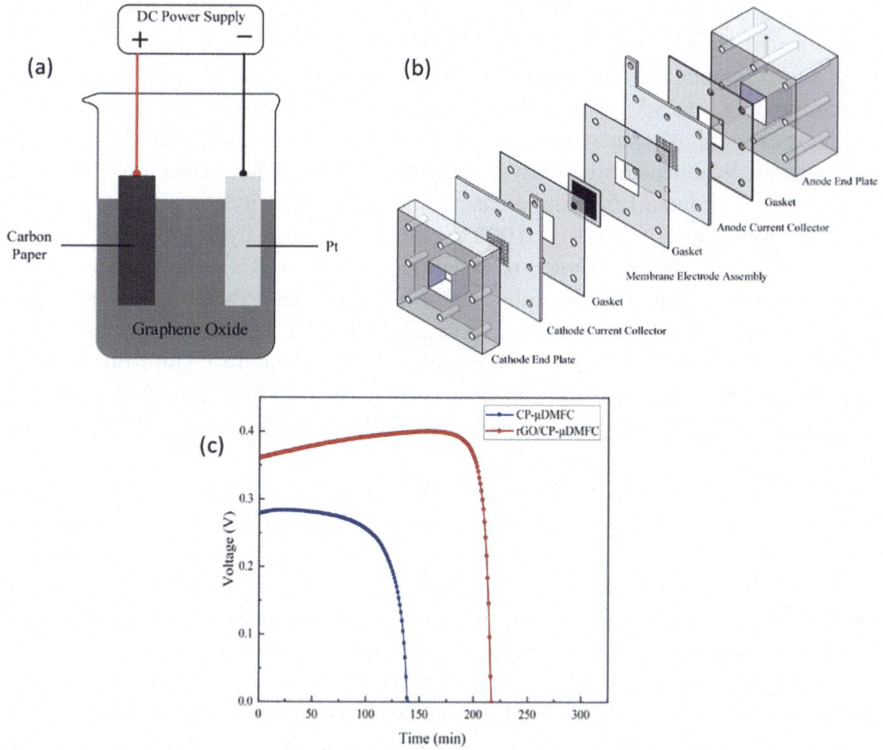

Fig. 4.7 Illustrate (**a**) carbon-based SOFC, (**b**) disassembly of the designed microDMFC, and (**c**) comparison graph between voltage and time for different materials [34]

instance, a membrane of polymer electrodes was designed and developed using zeolite and Chitosan incorporating rGO. As the rGO fraction within the Chitosan-zeolite/rGO membrane increased, the proton conductivity increased and reached a maximum value of 2 wt.% with electrolyte conductivity of 6.777×10^{-6} S/cm. It also exhibited an optimal value for ion exchange capacity of 0.18121 mmol g^{-1} [35]. The effective utilisation of expensive Pt-group metals through the aid of co-catalytic activity support is necessary for the commercialisation of DMFCs. A durable electrical connection improves the membrane's active sites and improves the transfer of charge, and a wide surface area was made possible by the MoWC/rGO composite in addition to Pt nanoparticles. With mass activity as well as area activity of 1486.1/641.7 mA mg^{-1} Pt and 60.11/25.94 mA cm^{-2}, onset and half-wave potentials of 1.12/1.02 V and 0.88/0.90 V, as well as exceptional stability and anti-poisoning qualities, the optimised Pt/MoWC/rGO catalyst outperforms Pt on Vulcan XC-72 carbon due to its superior MOR and ORR performance in both acidic and alkaline media [36].

A 3D iron and nickel-functionalised nitrogen and phosphorus-doped graphene foam (FeNi@NP-GF) was synthesised through a single-step pyrolysis process as a versatile electrocatalyst. For the oxygen reduction reaction (ORR), FeNi@NP-GF outperformed NP-GF, Fe@NP-GF, and Ni@NP-GF in terms of performance, stability, and tolerance to methanol. Notably, its onset potential (−0.06 V) and half-wave potential (−0.23 V) were 100 mV and 90 mV more positive than commercial Pt/C, respectively. The incorporation of the FeNi alloy significantly boosted the ORR activity. When anchored with the platinum for methanol oxidation, Pt/FeNi@NP-GF demonstrated a 1.6-fold increase in the current density compared to the Pt/C and exceptional stability for over 1000 cyclic voltammetry cycles [37]. In terms of electrocatalytic attributes, PtNiCo/rGO nanocomposites (NCs), which are produced via a microwave-assisted technique, perform better than PtNiCo catalysts. These include greater methanol oxidation activity, increased CO tolerance, superior stability, and a higher ECSA (87.41 m^2/g at 50 mV/s). These benefits result from the huge surface area, conductivity, and uniform dispersion of 17.03 ± 1.93 nm PtNiCo nanoparticles in rGO. The findings point to PtNiCo/rGO NCs as a viable substitute for anode catalysts in fuel cells [38].

A membrane-electrode structure with an optimised graphene-based interlayer between the anode catalyst and membrane enhances DMFC performance with high-concentration methanol fuel was designed in a research study. The maximum power density for the designed membrane reached 17.75 mW cm^{-2} under the influence of 8 M methanol when a Nafion ionomer binder was added to the layer of the graphene. The increment is 12.36% higher than a normal electrode and 82.69% higher for an electrode with no binder. Anode ECSA has increased by 40.16%, and the resistance for charge transfer has decreased by five times, according to the EIS and CV test. This improvement increases the energy density by improving DMFC's overall performance with high-concentration methanol [39]. In another strategy, spherical cerium oxide (CeO_2) and cerium-zirconium oxide (CeO_2–ZrO_2) solid solutions were made by hydrothermal synthesis. The Pt/CeO_2–ZrO_2/GO and Pt/CeO_2/GO catalysts were then made by supporting these oxides on graphene to evaluate the effect of the methanol on the stability of anode in fuel cell, electrocatalytic performance, and CO tolerance. The finding of the study revealed that the developed catalyst Pt/CeO_2–ZrO_2/GO exhibited strong catalytic performance along with high resilience towards the poisoning [40].

4.5 Graphene for Solid Oxide Fuel Cells

A potential high-temperature energy conversion technology that provides effective and sustainable power production is solid oxide fuel cells (SOFCs). However, especially at lower operating temperatures, the cathode performance may restrict the efficiency of these fuel cells. SOFC utilises an oxide material electrolyte in the solid state and is regarded as a simple, gas-to-solid method. Unlike liquid electrolytes, it shows no leakage issue, flooding of catalyst or even a poor ORR. However, for effective functioning, suitable material exhibiting high chemical and thermal stability is a challenge to find. The addition of graphene to solid oxides has been shown to improve the structural stability, electronic conduction, and redox reaction kinetics of the cathode material, leading to enhanced electrochemical performance [41].

Numerous strategies, including flat plate to tubular cathodes, have been applied for the development of SOFCs. However, these designs have their limitations, as flat plates may stack, and tubular ones can have sealing issues. Different chamber designs, including monolithic structures or single chambers, have also been studied for enhancing the performance of the SOFCs. For example, a high-efficiency system that combines a solid oxide fuel cell with a graphene thermionic energy converter was evaluated. The obtained result confirmed its possible application in SOFC with high-power density [42]. Similarly, SOFC exhausts heat is used to produce additional power in a hybrid system that combines an SOFC with a graphene-collector thermionic energy converter (GTEC). The system outperforms solo SOFCs by a factor of 1.2, reaching a maximum power density of 0.774 W cm^{-2} at 1073 K. To increase system efficiency, finite-time thermodynamics analysis determines the best operating conditions and important design parameters, such as area ratio, higher SOFC temperatures, better heat transfer, lower thermal emissivity, and optimised optical reflectors [43].

Graphene added to the electrodes can have a significant impact in reducing the operating temperature for SOFCs. For example, GO is included in $La_{0.3}Sr_{0.7}Fe_{0.4}Ti_{0.6}O_3$-δ to construct SOFC cathode material using a sol-gel process. Utilising the air, an electrical conductivity test was carried out between 300 and 600 °C. At 600 °C, GO-incorporated cathode material demonstrated improved electrical performance with conductivity reaching 7 S cm^{-1} for 1.5 wt% GO-$La_{0.3}Sr_{0.7}Fe_{0.4}Ti_{0.6}O_3$-δ electrode material. The composite with the same composition also exhibited the lowest area-specific resistance (ASR) of just 0.02 Ω cm^{-2} with a power density having a maximum value reaching 362 mW cm^{-2} [44]. The Hf-doped $SrFe_{0.9}Hf_{0.1}O_3$-δ (SFHf) perovskite is a Co-free cathode that exhibits improved chemical stability and ORR activity. In an anode-supported fuel cell, it shows a high-power density (1.94 W·cm^2 at 700 °C) and low polarisation resistance (0.193 Ω·cm^2 at 600 °C). Because of its high oxygen vacancy concentration, electrical conductivity, and quick oxygen kinetics, SFHf exhibits consistent ORR performance over 120 h in both air and CO_2. By strengthening metal-oxygen bonds and lowering oxygen vacancies, Hf doping increases chemical stability [45]. In SOFCs, perovskite-type oxides, as cathode material, have been widely investigated in the past decade. These materials offer superior traits in the temperature range from 500 to 800 °C [46]. The SOFCs have shown considerable enhancement in exhibiting the high-power density which occurs at the proximity of the cathode due to ORR. For example, it was observed that $La_{0.6}Sr_{0.4}Al_{0.2}Cu_{0.8}O_3$ added with graphene has been more thermally stable due to the mesoporous morphology. The electrical behaviour of the cathode measured using a four-probe device was 78.7 S cm^{-1} at a temperature ranging from 500 °C, which. Moreover, the open circuitry voltage reached a maximum value of 1.024 V, along with a power density of 513 mW cm^{-2} at a temperature value of 550 °C [47]. To improve the SOF cell performance, reduced graphene oxide (rGO) was added to the cathode functional layer (CFL) to get the advantages of the superior electrical characteristics of graphene. The schematics of SOFC

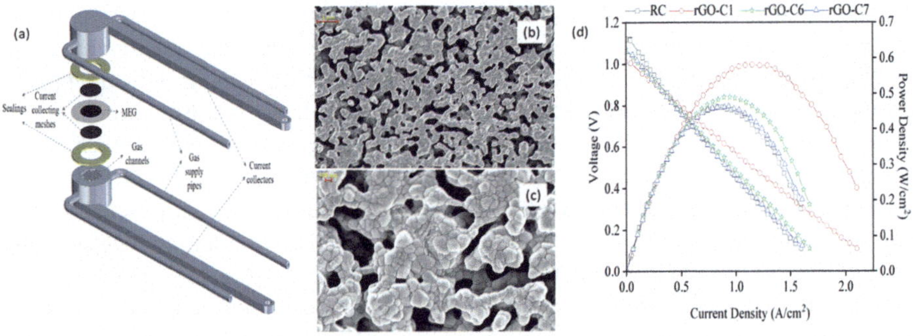

Fig. 4.8 (**a**) Represent the structure of SOFC stack (**b**) and (**c**) compare the microstructure of CFL and graphene-based CFL, and (**d**) show the relation between voltage, power density, and current density for different samples of CFL and rGO-based CFL [48]

stacking, microstructural images, and the power density of the different compositions are exhibited in Fig. 4.8. Electrochemical investigation from the developed samples reveals a ~ 26% increase in peak performance with a single layer of CFL, which contains 1 wt% graphene and sintered at a high temperature of 1050 °C [48].

As more advanced materials for electrodes for energy conversion devices, mixed metallic oxides are tempting researchers. At very low temperatures, the electrical conductivity of $Li_{0.1}Cu_{0.2}Zn_{0.7}$-oxide (LCZ), as well as the fuel cell performance, is significantly improved by the addition of 1.5% graphene. The graphene-LCZ material, which was synthesised using an economical solid-state method, has a low area-specific resistance (ASR) of 0.057 Ωcm^2 at a temperature condition of 450 °C in air and a conductivity of 3.5 S cm^{-2}. With hydrogen as fuel, a three-layer fuel cell with NSDC as the electrolyte exhibits a peak power density of 267.5 mW/cm^2 and an open-circuit voltage of 0.95 V [49].

4.6 Graphene Catalysts—A Leap Forward

Supporting platinum-based catalysts is one of the main uses of graphene in fuel cells. One utmost step in fuel cells is the ORR, for which platinum is considered highly. However, its poor durability and cost have restricted its commercialisation effectively. On the other hand, the quantity of Pt required can be decreased substantially by preserving or even increasing the catalytic activity by employing graphene as an added substrate. Platinum NPs may also scatter widely over the vast graphene surface area, which increases their use and enhances mass transfer activity. Furthermore, by limiting the sintering and deactivation, the strong connection between platinum and graphene might improve the stability of the catalyst. The creation of metal-free catalysts is another domain which can make exciting use of graphene in fuel cells. To construct active sites for the ORR, graphene can be functionalised with heteroatoms like sulphur, boron, or nitrogen. Compared to conventional metal-based catalysts, these graphene-based catalysts provide innumerable benefits, such as low cost, high stability, and poisoning resistance. Furthermore, by adjusting the type and concentration of dopants with graphene, they may be tailored for specific applications.

Graphene may be utilised as a component of membrane electrode assembly (MEA) in addition to acting merely as a catalyst support. In contrast to conventional polymer-based membranes, graphene-based membranes have shown better mechanical strength, chemical stability, and proton conductivity. Graphene can be added to the MEA to improve the fuel cell's overall durability and efficiency.

4.7 Conclusion—Graphene's Bright Horizon

Materials based on graphene have great potential to improve fuel cell performance significantly and efficiently. Graphene can be added to catalysts and membranes in proton exchange membrane fuel cells (PEMFCs) to increase their stability, catalytic activity, and proton conductivity. Higher power, rapid reaction times, and longer lifespans would be expected with the addition of graphene into the PEMFCs as catalyst support. Graphene-based materials can be employed as cathode materials in solid oxide fuel cells (SOFCs) to increase oxygen reduction kinetics, which would lower operating temperatures and improve cell performance. Additionally, the development of advanced fuel cell systems, such as direct methanol fuel cells (DMFCs), shows great benefits from the application of graphene-based materials. Graphene can aid in the creation of clean and sustainable energy solutions by enhancing the robustness and efficiency of these devices. Numerous fascinating paths for further study and invention are sprouting up as investigations into the potential of graphene in fuel cells continue. In forming cutting-edge fuel cell systems with multifunctional materials and 3D hierarchical structures utilising the benefits of graphene. Creating innovative catalysts for a range of fuel cell processes that have improved stability, selectivity, and activity. Utilising the potential of graphene can further pave the way for the development of novel fuel cell technology.

References

1. M. Bilgili, Y.E. Gönülaçar, Investigation of the effect of operating parameters on NERNST voltage in hydrogen-oxygen fuel cells. Isı Bilimi ve Tekniği Dergisi **44**(1), 59–69 (2024). https://doi.org/10.47480/isibted.1494033
2. M.S. Guney, Y. Tepe, Classification and assessment of energy storage systems. Renew. Sust. Energ. Rev. **75**, 1187–1197 (2017). https://doi.org/10.1016/j.rser.2016.11.102
3. S. Mitsushima, B. Gollas, and V. Hacker, "Introduction," in Fuel Cells and Hydrogen, Elsevier, 2018, pp. 1–13. doi: https://doi.org/10.1016/B978-0-12-811459-9.00001-3
4. H. Tateishi et al., Graphene oxide fuel cell. J. Electrochem. Soc. **160**(11), F1175–F1178 (2013). https://doi.org/10.1149/2.008311jes

5. T. Bayer, R. Selyanchyn, S. Fujikawa, K. Sasaki, S.M. Lyth, Spray-painted graphene oxide membrane fuel cells. J Memb Sci **541**, 347–357 (2017). https://doi.org/10.1016/j.memsci.2017.07.012
6. H. Alnaqbi, E.T. Sayed, S. Al-Asheh, A. Bahaa, H. Alawadhi, M.A. Abdelkareem, Current progression in graphene-based membranes for low temperature fuel cells. Int. J. Hydrog. Energy **52**, 800–842 (2024). https://doi.org/10.1016/j.ijhydene.2022.03.255
7. Z. Du, F. Yu, J. Wang, J. Li, X. Wang, A. Qian, Catalytic effects of graphene structures on Pt/graphene catalysts. RSC Adv. **14**(31), 22486–22496 (2024). https://doi.org/10.1039/D4RA02841D
8. C. Sun, Q. Zuo, G. Hu, Y. Xia, Graphene composite coating for enhanced corrosion resistance of Ni foam flow field in PEMFC. J. Alloys Compd. **1010**, 177251 (2025). https://doi.org/10.1016/j.jallcom.2024.177251
9. Y. Xia, Q. Zuo, C. Sun, G. Hu, B. Fang, Hierarchically electrodeposited nickel/graphene coatings for improved corrosion resistance of Ni foam flow field in PEMFC. Inorganics (Basel) **12**(11), 293 (2024). https://doi.org/10.3390/inorganics12110293
10. R. Kumar, K.A. Subramanian, Performance enhancement of hydrogen PEM fuel cell using graphene coated graphite electrodes. Process Saf. Environ. Prot. **190**, 48–59 (2024). https://doi.org/10.1016/j.psep.2024.07.031
11. Q. Yin, X.-W. Wang, S. Liu, X.-Z. Wang, X.-Z. Fu, J.-L. Luo, High corrosion resistance of reduced graphene oxide coated 316L stainless steel bipolar plate for proton exchange membrane fuel cell prepared by a facile method. Mater. Chem. Phys. **290**, 126663 (2022). https://doi.org/10.1016/j.matchemphys.2022.126663
12. M. Yaldagard, M. Arkas, Enhanced mass activity and durability of bimetallic Pt-Pd nanoparticles on sulfated-zirconia-doped graphene Nanoplates for oxygen reduction reaction in proton exchange membrane fuel cell applications. Molecules **29**(9), 2129 (2024). https://doi.org/10.3390/molecules29092129
13. S. Dhali et al., Waste plastic derived nitrogen-doped reduced graphene oxide decorated core–shell nano-structured metal catalyst (WpNrGO-Pd–Ru) for a proton exchange membrane fuel cell. Mater Adv **5**(9), 3771–3782 (2024). https://doi.org/10.1039/D3MA01006F
14. N. Manna, M. Singh, S. Kurungot, Microporous 3D-structured hierarchically entangled graphene-supported Pt 3 co alloy catalyst for PEMFC application with process-friendly features. ACS Appl. Mater. Interfaces **15**(23), 28023–28035 (2023). https://doi.org/10.1021/acsami.3c03372
15. W. Zhu et al., Space confinement to regulate ultrafine CoPt nanoalloy for reliable oxygen reduction reaction catalyst in PEMFC. Adv Science **10**(19) (2023). https://doi.org/10.1002/advs.202206062
16. S. Hu, Z.-A. Li, Facile synthesis of PtCo nanoparticles/three-dimensional graphene hybrid material as a highly active and stable electrocatalyst for oxygen reduction reaction. Chem. Eng. J. **471**, 144828 (2023). https://doi.org/10.1016/j.cej.2023.144828
17. Z. Gao et al., Pt nanoparticles supported on iron and nitrogen-doped holey graphene for boosting ORR performance. ACS Appl Nano Mater **6**(12), 10521–10530 (2023). https://doi.org/10.1021/acsanm.3c01502
18. M. Beltrán-Gastélum et al., Pt-Au nanoparticles on graphene for oxygen reduction reaction: Stability and performance on proton exchange membrane fuel cell. Energy **181**, 1225–1234 (2019). https://doi.org/10.1016/j.energy.2019.06.033
19. G.U. Alpaydin, Y. Devrim, C.O. Colpan, Performance of an HT-PEMFC having a catalyst with graphene and multiwalled carbon nanotube support. Int. J. Energy Res. **43**(8), 3578–3589 (2019). https://doi.org/10.1002/er.4504
20. R.S. Mane, A.S. Nair, R.I. Jafri, N. Jha, Zeolite framework-anchored carbon-doped white graphene as antipoisoning cathode materials for proton-exchange membrane fuel cells. ACS Appl Nano Mater **7**(20), 23454–23465 (2024). https://doi.org/10.1021/acsanm.4c01704

21. M. Akin, M. Bekmezci, R. Bayat, I. Isik, F. Sen, Ultralight covalent organic frame graphene aerogels modified platinum-magnetite nanostructure for direct methanol fuel cell. Fuel **357**, 129771 (2024). https://doi.org/10.1016/j.fuel.2023.129771
22. L. Guan et al., Effect of graphene aerogel as a catalyst layer additive on performance of direct methanol fuel cell. Fuel **360**, 130503 (2024). https://doi.org/10.1016/j.fuel.2023.130503
23. J. Xiong, Q. Zhang, H. He, H. Huang, Ultrasmall Rh-decorated porous heterostructures stereo-assembled from MXene and graphene for efficient methanol electrooxidation. Int. J. Hydrog. Energy **92**, 535–543 (2024). https://doi.org/10.1016/j.ijhydene.2024.10.271
24. D. Taechaboonsermsak et al., Calcium hydroxy phosphate-functionalized graphene oxide/Nafion composite membrane for direct methanol fuel cell. Int. J. Hydrog. Energy **52**, 1093–1110 (2024). https://doi.org/10.1016/j.ijhydene.2023.09.069
25. Y. Wang et al., Nickel/graphene/polyaniline composites prepared by one-step electrodeposition as catalysts for methanol oxidation. React. Kinet. Mech. Catal. **137**(4), 1871–1884 (2024). https://doi.org/10.1007/s11144-024-02621-1
26. V. Erduran, R. Bayat, I. Isik, T. Bayazit, F. Şen, Production of sustainable low-layer graphene by green synthesis at room conditions for platinum-based direct methanol fuel cell. ACS Engineering Au **3**(6), 537–545 (2023). https://doi.org/10.1021/acsengineeringau.3c00040
27. A.A. Shalaby, A.N. Aziz, Z. Špitalský, A.M. Omer, M.S. Mohy-Eldin, R.E. Khalifa, An effective methanol-blocking cation exchange membrane modified with graphene oxide Nanosheet for direct methanol fuel cells. PRO **11**(2), 353 (2023). https://doi.org/10.3390/pr11020353
28. W.W. Ng, H.S. Thiam, Y.L. Pang, Y.S. Lim, J. Wong, L.H. Saw, Self-sustainable, self-healable sulfonated graphene oxide incorporated nafion/poly(vinyl alcohol) proton exchange membrane for direct methanol fuel cell applications. J. Environ. Chem. Eng. **11**(6), 111151 (2023). https://doi.org/10.1016/j.jece.2023.111151
29. N. Metzger, I. Vlassiouk, S. Smirnov, G. Mariscal, R. Spragg, X. Li, Experimental studies of graphene-coated polymer electrolyte membranes for direct methanol fuel cells. J Electrochem Energy Convers Storage **20**(2) (2023). https://doi.org/10.1115/1.4056269
30. S.H. Osman, S.K. Kamarudin, S. Basri, N.A. Karim, Anodic catalyst support via titanium dioxide-graphene aerogel (TiO2-GA) for a direct methanol fuel cell: Response surface approach. Catalysts **13**(6), 1001 (2023). https://doi.org/10.3390/catal13061001
31. S.H. Osman, S.K. Kamarudin, S. Basri, N.A. Karim, Potential of 3D hierarchical porous TiO2-graphene aerogel (TiO2-GA) as electrocatalyst support for direct methanol fuel cells. Nano **13**(12), 1819 (2023). https://doi.org/10.3390/nano13121819
32. S.A. Gokulakrishnan, V. Kumar, G. Arthanareeswaran, A.F. Ismail, J. Jaafar, Thermally stable nanoclay and functionalized graphene oxide integrated SPEEK nanocomposite membranes for direct methanol fuel cell application. Fuel **329**, 125407 (2022). https://doi.org/10.1016/j.fuel.2022.125407
33. X. Liu et al., Efficient and durable composite membranes based on polydopamine-mediated sulfonated graphene oxide for direct methanol fuel cells. J Mater Chem A Mater **9**(42), 24044–24055 (2021). https://doi.org/10.1039/D1TA07882H
34. D. Zhang, K. Li, Z. Wang, Z. Zhao, Reduced graphene oxide/carbon paper for the anode diffusion layer of a micro direct methanol fuel cell. Nano **12**(17), 2941 (2022). https://doi.org/10.3390/nano12172941
35. Y.A. Sihombing, Susilawati, S.U. Rahayu, M.D. Situmeang, Effect of reduced graphene oxide (rGO) in chitosan/Pahae natural zeolite-based polymer electrolyte membranes for direct methanol fuel cell (DMFC) applications. Mater Sci Energy Technol **6**, 252–259 (2023). https://doi.org/10.1016/j.mset.2023.01.002
36. S. Li, Y. Zhang, Y. Han, F. Lv, B. Liu, L. Huo, Bimetallic molybdenum-tungsten carbide/reduced graphene oxide hybrid promoted Pt catalyst with enhanced electrocatalytic activity and stability for direct methanol fuel cells. Appl. Surf. Sci. **600**, 154134 (2022). https://doi.org/10.1016/j.apsusc.2022.154134

37. C. Shi, X. Maimaitiyiming, FeNi-functionalized 3D N, P doped graphene foam as a noble metal-free bifunctional electrocatalyst for direct methanol fuel cells. J. Alloys Compd. **867**, 158732 (2021). https://doi.org/10.1016/j.jallcom.2021.158732
38. K.-Y. Shih, J.-J. Wei, M.-C. Tsai, One-step microwave-assisted synthesis of PtNiCo/rGO electrocatalysts with high electrochemical performance for direct methanol fuel cells. Nano **11**(9), 2206 (2021). https://doi.org/10.3390/nano11092206
39. Q. Xu et al., Constructing a graphene-contained layer in anode to improve the performance of direct methanol fuel cells using high-concentration fuel. Int J Green Energy **18**(6), 566–577 (2021). https://doi.org/10.1080/15435075.2020.1865374
40. Q. Wang et al., Graphene and spherical CeO 2 -ZrO 2 support Pt nanoparticles as anode catalysts for direct methanol fuel cells. J. Electrochem. Soc. **168**(5), 052507 (2021). https://doi.org/10.1149/1945-7111/ac0305
41. Q. Liu et al., Graphene-modified nanostructured vanadium pentoxide hybrids with extraordinary electrochemical performance for Li-ion batteries. Nat. Commun. **6**(1), 6127 (2015). https://doi.org/10.1038/ncomms7127
42. T. Liao, Y. Dai, C. Cheng, Q. He, Q. Xu, M. Ni, Parametric optimization of a coupled system integrating solid oxide fuel cell and graphene thermionic energy converter. J. Power Sources **478**, 228797 (2020). https://doi.org/10.1016/j.jpowsour.2020.228797
43. A. Ding et al., Thermodynamic analysis and parameter optimization of a hybrid system based on SOFC and graphene-collector thermionic energy converter. Energy Convers. Manag. **291**, 117327 (2023). https://doi.org/10.1016/j.enconman.2023.117327
44. K. Ahmad, G. Abbas, M.A. Ahmad, Z. Kazim, T. Munir, Preparation and characterization of nanocomposite perovskite cathode materials La 0.3 Sr 0.7 Fe 0.4 Ti 0.6 O 3-δ (LSFT) for low-temperature SOFCs with incorporation of graphene oxide (GO). Cryst. Res. Technol. **59**(1) (2024). https://doi.org/10.1002/crat.202300197
45. Z. Zhang, D. Chen, J. Wang, S. Tan, X. Yu, Z. Shao, Highly active and stable cobalt-free hafnium-doped SrFe 0.9 Hf 0.1 O 3−δ perovskite cathode for solid oxide fuel cells. ACS Appl Energy Mater **1**(5), 2134–2142 (2018). https://doi.org/10.1021/acsaem.8b00198
46. Z. Tao et al., High-performing proton-conducting solid oxide fuel cells with triple-conducting cathode of Pr0.5Ba0.5(Co0.7Fe0.3)O3-δ tailored with W. Int. J. Hydrog. Energy **47**(3), 1947–1953 (2022). https://doi.org/10.1016/j.ijhydene.2021.10.145
47. S.A. Memon, H. Shaikh, R. Raza, Z.U.N. Mughal, A.A. Memon, S. Memon, Graphene incorporated mesoporous perovskite with excellent conductivity and catalytic activity for low temperature solid oxide fuel cells. New J. Chem. **46**(26), 12530–12539 (2022). https://doi.org/10.1039/D2NJ01917E
48. C. Timurkutluk et al., Effect of reduced graphene oxide addition on cathode functional layer performance in solid oxide fuel cells. Int. J. Hydrog. Energy **48**(60), 23127–23135 (2023). https://doi.org/10.1016/j.ijhydene.2022.11.082
49. M.A. Ahmad et al., Synergistic electrochemical properties of graphene incorporated LCZ-oxide cathode for low temperature solid oxide fuel cell. Crystals (Basel) **13**(3), 434 (2023). https://doi.org/10.3390/cryst13030434

Index

V. Khanna, C. Prakash, *Graphene for Electrochemical Energy Storage*, Lecture Notes in Nanoscale Science and Technology 24,
https://doi.org/10.1007/978-3-031-87643-1

The manufacturer's authorised representative in the EU is Springer Nature Customer Service Centre GmbH, Europaplatz 3, 69115 Heidelberg, Germany. If you have any concerns regarding our products, please contact ProductSafety@springernature.com

Printed and bound by CPI Group (UK) Ltd, Croydon, CR0 4YY
30/09/2026
02225607-0003